M000237976

Cases in Clinical Infectious Disease Practice

Cases in Clinical Infectious Disease Practice

Obtaining a good history from the patient remains the cornerstone of an accurate clinical diagnosis: Lessons learned in many years of clinical practice

Okechukwu Ekenna

WILEY Blackwell

Copyright © 2016 by John Wiley & Sons, Inc. All rights reserved

Published by John Wiley & Sons, Inc., Hoboken, New Jersey
Published simultaneously in Canada

No part of this publication may be reproduced, stored in a retrieval system, or transmitted in any form or by any means, electronic, mechanical, photocopying, recording, scanning, or otherwise, except as permitted under Section 17 or 18 of the 1976 United States Copyright Act, without either the prior written permission of the Publisher, or authorization through payment of the appropriate per-copy fee to the Copyright Clearance Center, Inc., 222 Rosewood Drive, Danvers, MA 1923, (978) 750-8400, fax (978) 750-4470, or on the web at www.copyright.com. Requests to the Publisher for permission should be addressed to the Permissions Department, John Wiley & Sons, Inc., 111 River Street, Hoboken, NJ 730, (21) 748-611, fax (21) 748-608, or online at http://www.wiley.com/go/permission.

Limit of Liability/Disclaimer of Warranty: While the publisher and author have used their best efforts in preparing this book, they make no representations or warranties with respect to the accuracy or completeness of the contents of this book and specifically disclaim any implied warranties of merchantability or fitness for a particular purpose. No warranty may be created or extended by sales representatives or written sales materials. The advice and strategies contained herein may not be suitable for your situation. You should consult with a professional where appropriate. Neither the publisher nor author shall be liable for any loss of profit or any other commercial damages, including but not limited to special, incidental, consequential, or other damages.

For general information on our other products and services or for technical support, please contact our Customer Care Department within the United States at (800) 762-2974, outside the United States at (317) 572-3993 or fax (317) 572-402.

Wiley also publishes its books in a variety of electronic formats. Some content that appears in print may not be available in electronic formats. For more information about Wiley products, visit our web site at www.wiley.com.

Library of Congress Cataloging-in-Publication Data:

Names: Ekenna, Okechukwu, author.
Title: Cases in clinical infectious disease practice / Okechukwu Ekenna.
Description: Hoboken, New Jersey : John Wiley & Sons, Inc., [2016] | Includes
  bibliographical references and index.
Identifiers: LCCN 2016023166 (print) | LCCN 2016023807 (ebook) | ISBN
  9781119044161 (cloth) | ISBN 9781119044185 (pdf) | ISBN 9781119044062
  (epub)
Subjects: | MESH: Communicable Diseases–diagnosis | Communicable
  Diseases–therapy | Clinical Medicine–methods | Case Reports
Classification: LCC RC112 (print) | LCC RC112 (ebook) | NLM WC 100 | DDC
  616.9–dc23
LC record available at https://lccn.loc.gov/2016023166

Printed and bound in Malaysia by Vivar Printing Sdn Bhd

10 9 8 7 6 5 4 3 2 1

Contents

Table of Normal Laboratory Values*

Component	Unit	Reference range
White blood cell	K/UL	4.4–10.1 (4400–10,100/µL)
Hemoglobin	g/dL	10.0–18.0
Hematocrit	%	37–50
Platelets	K/UL	117–369
MCV	fL	82–99
Neutrophils (polymorphonuclear leukocytes)	%	43.7–84.9
Eosinophils	%	0.6–6.0
Lymphocytes	%	8.4–40.7
Erythrocyte sedimentation rate	mm/hour	0–20
C-reactive protein	mg/dL	0.00–0.30
O_2 saturation	%	95–99 (arterial/capillary); >74 (venous)
Glucose (fasting)	mg/dL	65–99
Blood urea nitrogen	mg/dL	7–18
Creatinine	mg/dL	0.6–1.3
Sodium	mmol/L	135–148
Potassium	mmol/L	3.5–5.3
Alanine aminotransferase (serum glutamic pyruvic transaminase)	IU/L	12–78
Aspartate aminotransferase (serum glutamic oxalo-acetic transaminase)	IU/L	15–37
Albumin	g/dL	3.4–5.0
Alkaline phosphatase	IU/L	45–117
Total bilirubin	mg/dL	0.2–1.0
Lactate dehydrogenase	IU/L	84–246

*Adapted from the 2015 Singing River Health System Adult Normal Laboratory Values. Variations with age and sex are not reflected here. These values are to be used only as a general guide.

Okechukwu Ekenna, MD, MPH, D(ABMM), FACP

Dr Ekenna was born in Aba, Nigeria, where he had his primary school and most of his secondary school education (Dennis Memorial Grammar School in Onitsha). He spent a year as an exchange student at Loughborough Grammar School in Leicestershire, England, and another year in Germany, at Neusprachliches Gymnasium Aue, in Wuppertal-Elberfeld. General Certificate of Education (GCE, Advanced Level) was acquired following attendance at the Modern Tutorial College in London.

He attended medical school at Philipps-Universitaet Marburg/Lahn, Germany, graduating in April, 1979 (Aerztliche Pruefung). He successfully defended his doctoral thesis in June, 1979 (magna cum laude) and spent a year of medical internship in Germany before moving to the United States for Residency and Fellowship trainings in Internal Medicine, Infectious Diseases,and Post-Doctoral Fellowship in Medical Microbiology.

Through the course of his clinical and teaching experiences, he has served as consultant physician and lecturer to several institutions in Nigeria (University of Maiduguri Teaching Hospital) and in the USA. He is presently Adjunct Associate Professor of Medicine (since 1998)at the University of South Alabama, in Mobile, AL. He is in private practice in Pascagoula, MS,and is Consultant in Infectious Diseases to the Singing River Health System (SRHS). He is the Chairman of the Infection Prevention/Control Committee for SRHS. He has active consulting privileges at Ocean Springs Hospital and Singing River Hospital in Pascagoula, MS.

He is a naturalized citizen of the United States.

Dr Ekenna is a member of the American Society for Microbiology, American Society of Tropical Medicine and Hygiene, Infectious Diseases Society of America, and Emerging Infections Network of the Infectious Diseases Society of America. He is a Fellow of the American College of Physicians and Fellow of the Royal Society of Tropical Medicine and Hygiene, London. He is also a member of the Organizing Committee of the African Initiative Group of the American Society for Microbiology.

Dr Ekenna is Board certified in Internal Medicine, Infectious Diseases, and in Medical and Public Health Microbiology. He is also certified in Clinical Tropical Medicine and Travelers' Health. He has a Master's degree in Public Health (MPH). He has authored multiple papers in peer-reviewed journals.

Dr Ekenna is fluent in Igbo (Nigerian), English, German, moderately so in French, and speaks a little Hausa. He is still trying to learn Spanish. His wife thinks of him as a history buff.

Dedication

To my parents: Eze Raymond Onuoha Ekenna and Ugo-eze (Mrs) Gabraeline Urasi Ekenna (née Ezurike), and their recognition of the value of a good education.

Acknowledgments

First, I would like to thank the many patients presented here (who will remain anonymous), who have allowed me to tell their story and use their image to illustrate the clinical points made in this book. We are reminded every day that it is a privilege to care for patients who entrust their lives and well-being to our judgments, often working with incomplete or conflicting data. Many of the presented images have made it easier to tell the stories in this book.

I am grateful to the colleagues who asked me to see their patients and render an opinion. I thank them for their trust, and the confidence reposed in me. It is clear that sometimes there are no clear answers to the complex questions posed by these physicians and their patients.

I also thank the many journal and book publishers who gave us permission to use data and figures from their publications and resources. We have acknowledged them wherever indicated. Those illustrations have served to improve the presentation of the cases in this book.

The Singing River Health System gave permission and approval to use laboratory and demographic data, as well as radiologic images where applicable, without patient identifiers or infringing on patient confidentiality. I am grateful for their support.

Ms Cyndi Aycock, MLT (senior microbiologist), was particularly helpful in working with me to process and preserve many microbiological cultures and specimens over the years, many of which are depicted in this book. She did this with such enthusiasm. Thank you, Cyndi.

Dr Sid Eudy, pathologist, was gracious enough to prepare some of the micrographs used in this book.

Several of my friends and colleagues reviewed this manuscript and offered useful suggestions: John O. Chikwem, PhD, Professor of Microbiology and Immunology, Lincoln University, Lincoln, Pennsylvania; Keith Ramsey, MD, Professor of Medicine, East Carolina University, Greenville, NC, formerly Head of Infectious Diseases at the University of South Alabama (USA) in Mobile, AL, who first recruited me as adjunct faculty at USA; and Abraham Verghese, MD, Professor of Medicine and Vice Chair for the Theory and Practice of Medicine, Stanford University, CA.

Dr Christopher Paddock, of the Centers for Disease Control and Prevention, was kind enough to write a generous preface for me. I thank him immensely.

Ms Cynthia Davis, my office nurse, was very helpful in tracking down some of the patients, so that I could obtain their formal permissions to use their photographs in the book.

I am grateful to John Wiley & Sons, Inc., my publishers, and especially to Ms Mindy Okura-Marszycki and Ms Stephanie Dollan, for trusting me initially with this opportunity; to my project editor, Divya Narayanan, and Anandhavalli Namachivayam, production editor, for working with me intensely to get this book readied for publication on time.

The copy editor, Holly Regan-Jones did a wonderful job to help clear up ambiguities.

Finally, I would like to thank my wife, Chiazo, for her support and encouragement over the several years it took to put this book together. She encouraged me as I worked over the holidays to add small sections to the book, in between other clinical and family responsibilities.

Preface

In the spring of 2007, I received a call from a colleague at the Mississippi State Department of Health who informed me of a patient recently evaluated by an infectious diseases physician from the Gulf Coast. The patient had presented with fever, rash, and an eschar following the bite of a tick. At the time of the call, the patient was no longer symptomatic, for the physician had presumptively and correctly diagnosed the illness as rickettsial infection and prescribed the appropriate antibiotic; however, he was curious to know if there was a way to determine if this illness represented a recently recognized disease known as *Rickettsia parkeri* rickettsiosis. Remarkably, he had retained the tick removed by the patient. Working with the state health department, the tick was sent to our laboratory at the Centers for Disease Control and Prevention, where we determined that it was in fact infected by *R. parkeri*, closing the loop and establishing this little-known agent as the cause of the patient's illness.

This episode was my first interaction with Dr Ekenna, the astute infectious disease physician and author of this book. To place Dr Ekenna's diagnostic acumen in greater perspective, he diagnosed four more patients with *R. parkeri* rickettsiosis over the next several years, which is particularly remarkable when one considers that our laboratory at CDC has identified only about 40 patients with this disease in the entire United States since its discovery in 2002.

The case studies provided in this book are salient examples of Dr Ekenna's keen ability of medical detection, whereby the diagnosis and care of each patient are based on a careful history and physical examination. These fundamental processes represent the pillars of clinical diagnosis and are emphasized considerably during the training of medical students and residents; nonetheless, years of practice and experience are characteristically needed to hone and channel these skills to a level where they can be applied effectively and consistently. The cases described herein illuminate Dr Ekenna's deceptively simple and logical approach to each medical mystery in which he gathers pertinent data relating to person, place, and time and assembles a diagnosis and careful plan of treatment tailored to the complete circumstances of the individual patient. This truly is the art of medicine. In this context, we follow the diagnosis and successful treatment of a financially challenged patient with sporotrichosis, learn of the removal of carious teeth from a patient with *Actinomyces israelii* endocarditis, and observe the confirmation of a sulfa drug-induced hypersensitivity masquerading as sepsis in a long-suffering patient.

The breadth and scope of cases portrayed in this collection are fascinating. These include a wide spectrum of infectious conditions caused by various common and not so common infectious agents that include mycobacteria, fungi, helminths, spirochetes, and, of course, rickettsiae. Many of the infections are presented as syndromic or uniquely situational processes, such as a section on serious soft tissue infections caused by various *Vibrio*, *Staphylococcus*, and *Alcaligenes* species bacteria identified in patients along the Gulf Coast in the aftermath of Hurricane Katrina. Remarkably, these clinical vignettes originate not from the collective encounters by a large group of specialists at a tertiary care facility in a large metropolitan center, but rather from the professional experience of one doctor working at a small community hospital along the Mississippi Gulf Coast, and reflect the immeasurable good that a thorough and thoughtful physician can provide to the health and well-being of an entire region.

Christopher D. Paddock, MD, MPHTM
Centers for Disease Control and Prevention, Atlanta, Georgia

Introduction

In the era of cost cutting and lack of adequate health insurance for many patients, clinical skills and time spent with patients are not adequately compensated. Yet these dwindling and underpaid skills – good history taking, observation of and listening to patients, and physical examination – remain essential to making and reaching a complete and accurate diagnosis. Expensive laboratory and imaging diagnostics, while very relevant, should not replace these age-old skills that have served to enhance and maintain the doctor–patient relationship and human connection, a connection that is often necessary for healing.

The process of differential diagnosis is particularly relevant in infectious diseases, and still involves what we typically label as "the art and science of medicine." This process requires a skill that usually improves with clinical experience, and is not achieved by textbook reading alone.

I have had the privilege and requirement to present on a regular basis real cases seen in my private practice over the last 18 years to medical students, residents, fellows, and faculty as part of my teaching responsibilities with the University of South Alabama in Mobile, Alabama, as well as presenting to the local medical staff on the Gulf Coast of Mississippi. The positive feedback I have received from students, medical staff, and faculty has encouraged me to present some of these cases in this publication. The format chosen is similar to the way I have usually presented them at the teaching conferences, except that they will not be in PowerPoint and will be somewhat abbreviated because of space constraints.

These cases will provide an illustration of how the infectious disease clinician processes and integrates data to arrive at a diagnosis. This type of hands-on approach is not given adequate emphasis these days in our training programs.

The cases presented in this book will take the reader from the initial patient encounter, through the history and physical examination, to simple laboratory findings and stains, to a final diagnosis, in a way that is simple to follow and without the need to cram up a lot of other data.

The book is intended for the practicing clinician or student in clinical training. It should be useful to teaching hospitals involved in the training of medical residents and students of allied health institutions involved in clinical practice. An additional advantage is that the cases presented here do not reflect patients seen at tertiary institutions, but rather in the community setting. They reflect the type of cases or situations the resident or student is likely to encounter in the real world after training.

The cases presented in this book should be within the reach of the average practicing physician and medical resident in training. They should also help practitioners and students in allied health, who may work up clinical specimens referred by clinicians, in their understanding of the thinking of the consultant. The cases will include photographs, illustrations, and microbiological slides as applicable and available. I have added schematic diagrams on the few occasions where it was not possible to obtain permits for patient photographs.

It is hoped that reviewing these cases will enhance the integrated skills and critical thinking of the reader.

Finally, at the end of each case, I will discuss diagnostic aids or clues in the case and practical lessons learned.

How the book should be used and understood

The cases presented were seen between 1997 and 2015, an 18-year period, and include inpatients and outpatients.

The presentations reflect the time and period during which the patients (cases) were encountered. Diagnostic and treatment modalities, therefore, reflect those available at the time and place of care.

All of these cases were consultations provided to physicians practicing in the community setting (whether hospital based or in office practice).

All personal identifiers have been removed (including names of institutions where care was provided), in order to protect the identity of the patients.

However, the dates of patient encounters (consults) have been included for context, as well as the season of the year, as these may provide important epidemiologic clues to making a correct diagnosis.

My suggestion is that you first review the case history and then hazard a diagnosis before turning to the answer and discussion section.

At the end of each case presentation, we will address simple diagnostic clues and any lessons learned from the case.

CHAPTER 1

Skin and Soft Tissue Infections

Case 1.1   Soft tissue infection following traumatic aquatic exposure

In early September 2010, a 7-year-old boy suffered a laceration injury to the left calf during a recreational boat ride on a coastal river. He was brought to the emergency room (ER) within 2 hours of this accident, after initial cleansing first aid with saline in the field.

The wound was noted to be severe and deep, and was described as "partial degloving" by the ER physician. The boy's vital signs were stable. Due to the nature of the wound, an orthopedic surgeon was consulted and the patient was taken to surgery within 2 hours for wound washout and closure. X-rays of the leg showed no fractures.

Cefazolin and gentamicin were given preoperatively. The patient received 34 stitches to close the large (>10 cm) complex left calf laceration injury after extensive washout. No cultures were done.

One day later, infectious diseases consult was sought for outpatient oral antibiotic recommendations, in anticipation of discharge home later that day.

The boy's past medical history was unremarkable, except for hospital admission for symptoms of nausea and vomiting 1 year earlier, and ear tubes placed 1.5 years earlier.

His examination was unremarkable except for superficial abrasions on the lower abdomen and right upper arm, and the deep (now sutured) left calf laceration. Temperature was 100.5 °F, but other vital signs were stable.

A combination of trimethoprim/sulfamethoxazole and cefuroxime was recommended as oral antibiotics, and the patient was then discharged to outpatient follow-up.

Six days after discharge from the hospital, he was readmitted because of wound infection. He had failed to take the prescribed antibiotics because of severe nausea.

At surgery, no frank pus was found but serous old blood and drainage were noted. A swab of this drainage was stained and cultured.

Cases in Clinical Infectious Disease Practice: Obtaining a Good History from the Patient Remains the Cornerstone of an Accurate Clinical Diagnosis: Lessons Learned in Many Years of Clinical Practice, First Edition. Okechukwu Ekenna.
© 2016 John Wiley & Sons, Inc. Published 2016 by John Wiley & Sons, Inc.

The infectious diseases consultant was called (after surgery) to help with additional recommendations. He added ceftriaxone to the vancomycin already prescribed. Examination of the patient the next day found him to be comfortable, afebrile, and eating breakfast. His temperature was normal (97.8 °F). The abdominal skin abrasions were healing, but the left calf was wrapped up following the surgical incision and debridement (I&D) the day before.

Basic laboratory findings were normal (white blood cell [WBC] count was 7300/μL, platelet count 401,000/μL, and creatinine 0.5 mg/dL).

The gram stain of the serous fluid from the wound showed rare WBCs and no organisms. Twenty-four hours later, the culture was reported positive for a gram-negative rod (GNR).

The left calf wound was clean when inspected on day 4 post surgery (Fig. 1.1a).

• *What are the likely organisms in this patient (differential diagnoses) and why?*

The GNR was found to be oxidase positive, beta-hemolytic on sheep blood agar (BA), and the subculture showed luxuriant growth on all three agar media (BA, chocolate BA, and MacConkey agar) within 24 hours (Fig. 1.1b). It was noted later to be resistant to penicillin/ampicillin-like agents, including carbapenems, but sensitive to second- and third-generation cephalosporins, quinolones, and trimethoprim/sulfamethoxazole, as well as tetracyclines and aminoglycosides.

• *What is your new diagnosis?*

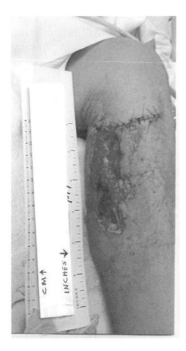

Figure 1.1a Left calf wound on day 4 post surgery (reproduced with permission). (*See insert for color information*)

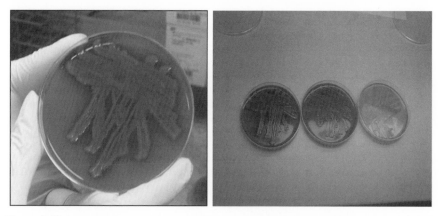

Figure 1.1b Luxuriant growth of gram-negative rod on blood agar medium in 24 hours, and comparative growth of the same organism on BA, chocolate BA, and MacConkey agar. (*See insert for color information*)

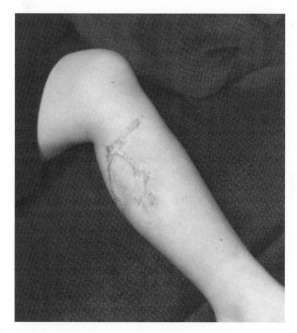

Figure 1.1c Photo taken in March 2011, 6 months after the injury: healed left calf post skin grafting (reproduced with permission).

The patient was discharged 6 days later to outpatient follow-up on intravenous ceftriaxone after a peripherally inserted central catheter (PICC line) was placed.

Two months later he was doing very well, and underwent plastic surgical repair of the calf laceration (Fig. 1.1c).

- *What were the clues to the diagnosis?*

Case discussion

First, the injury occurred while boating on a river, in possibly brackish water (coastal river). Gram-negative organisms are common and likely to contaminate the wound with such a severe laceration injury. The organism grew rapidly on all three culture media used (BA, Choc BA, and MAC). It was oxidase positive, non-lactose fermenting, mucoid, and beta-hemolytic on sheep BA. It did not require high salt concentration to grow, making certain *Vibrio* organisms (those that need high, 6.5% salt concentration) unlikely. The sensitivity pattern is also useful (see earlier). The organism turned out to be sensitive to the two agents originally recommended (trimethoprim/sulfamethoxazole and cefuroxime). The child was unable to take the antibiotics because of persistent nausea.

Differential diagnosis

Our differential diagnoses at the onset were *Aeromonas* or *Pleisiomonas*, more likely than *Vibrio*, because of the fresh water or brackish water environment where the injury occurred. The luxuriant growth in all three media was in keeping with the differential diagnoses chosen.

• *Final diagnosis: Aeromonas hydrophila*

Some data from the literature on *Aeromonas* are shown in Table 1.1a and given below.

Sites of isolation of *Aeromonas* (environmental reservoirs) [2]

• Fresh water
• Estuarine (brackish) water

Table 1.1a Organisms associated with soft tissue infection following water exposure.

Organisms	Exposure	Clinical syndromes
Aeromonas spp*	Fresh water	Rapidly developing infection associated with fever; sepsis
*Edwardsiella tarda**	Fresh water	Cellulitis, occasionally fulminant infection with bacteremia
Erysipelothrix rhusiopathiae	Puncture wounds from shrimp, crabs, and fish	Indolent localized cutaneous eruption; erysipeloid
*Vibrio vulnificus**	Salt or brackish water	Rapidly progressive necrotizing infection; bullous cellulitis; sepsis
Mycobacterium marinum	Salt or fresh water including fish tanks	Indolent infection; papules progressing to shallow ulcers; ascending lesions may resemble sporotrichosis

*Infection associated with patients with underlying liver disease, iron overload syndromes, and cancer; infections in these high-risk groups are particularly fulminant [1].
Reproduced with permission of UpToDate from Baddour [1].

- Surface water, especially recreational
- Drinking water, including treated, well, and bottled
- Polluted waters
- Waste water effluent sludge
  The organism grows at a range of temperatures from 0 °C to 42 °C.

Characterization and classification of *Aeromonas* [2]
- Aeromonads are ubiquitous inhabitants of fresh and brackish water.
- They have also been recovered from chlorinated tap water, including hospital water supplies.
- They occasionally cause soft tissue infections and sepsis in immunocompromised hosts and increasingly have been associated with diarrheal disease.
- Because of recent phylogenetic studies, *Aeromonas* species have been moved from the family Vibrionaceae to a new family, the Aeromonadaceae.

Microbiologic identification of *Aeromonas* [2,3]
- Oxidase-positive, polar flagella, glucose fermenter, facultative anaerobic GNR.
- Resistant to the vibriostatic agent O/129, and unable to grow in 6.5% NaCl.
- Hemolysis is variable but most are beta-hemolytic on BA media.
- There are a few useful standard biochemical tests (*A. hydrophila* is catalase positive and motile, converts nitrate to nitrite, and is urease negative).

Antimicrobial susceptibility tests of motile *Aeromonas* spp [2]
- Most strains are resistant to penicillin, ampicillin, carbenicillin, and ticarcillin.
- Most are susceptible to second- and third-generation cephalosporins, aminoglycosides, (carbapenems), chloramphenicol, tetracyclines, trimethoprim/sulfamethoxazole (TMP-SMX), and fluoroquinolones. In the case presented here, the *Aeromonas* was resistant to carbapenems, ampicillin, and first-generation cephalosporins.
- Higher resistance patterns (to TMP-SMX, tetracycline, and some extended-spectrum cephalosporins) have been found in Taiwan and Spain.
- Therapy with ampicillin or first-generation cephalosporins is not appropriate.

Lessons learned from this case
- Think of wound contamination with any complex laceration injury.
- Culture of the wound is critical before antibiotics are prescribed.
- No complete closure of wound, especially when environmental contamination is known or anticipated, as in this case.
- The patient should be advised to take the antibiotics or report if there are problems, so adjustments may be made. The patient in this case did not take the prescribed antibiotics because of nausea.
- Inspect wounds regularly for signs of inflammation or infection (pus, redness, pain, swelling, induration, or drainage).

- Think of the epidemiology (type of injury, environment, place, and season of the year) of the injury, and likely organisms expected in the given circumstances

References

1 Baddour LM.Soft tissue infections following water exposure. Available at: www.uptodate .com/contents/soft-tissue-infections-following-water-exposure?view=print#, accessed February 24, 2016.

2 Morris JG, Horneman A. Aeromonas infections. Available at: www.uptodate.com/contents/ aeromonas-infections?topicKey=ID%2F3138&elapsedTimeMs=5&source=search_result& searchTerm=Aeromonas+infections&selectedTitle=1%7E28&view=print&displayedView=full, accessed February 24, 2016.

3 Steinberg JP, Burd EM. Other gram-negative and gram-variable bacilli. In: Mandell GL, Bennett JE, Dolin R (eds) *Principles and Practice of Infectious Diseases*, 7th edn. Philadelphia: Churchill Livingstone, 2010, pp. 3017–19.

Case 1.2   Soft tissue infection of the hand and wrist in an 81-year-old man

An 81-year-old Caucasian male was seen for soft tissue infection of the right hand and wrist.

Infectious diseases (ID) consult was sought on 3/4/09 because of a severe hand and wrist wound following surgery. This patient had extensive flexor tenosynovectomy of the right wrist and drainage of abscess of the palm, plus carpal tunnel release 16 days earlier on 2/16/09.

He had been seen 6 weeks earlier in January elsewhere because of swelling of the right thumb several days after a shrimp fin puncture injury to his thumb. Following computed tomography (CT), magnetic resonance imaging (MRI), and ultrasound studies, he had been started on antibiotics without improvement in the swelling of the thumb. When he began to develop cellulitis of the right arm, he was referred to the orthopedic surgeon for further management. Initially, steroid injection to the right wrist was tried, along with oral clarithromycin, but these treatments did not help the swelling. Incision and debridement was therefore done on 2/16/09. At surgery, thick white, purulent material (pus) was obtained, and sent for culture.

The significant finding in the review of systems was severe pain in the wrist, but no fever.

Past medical history and underlying diseases included the following: Atrial fibrillation, for which he had been on Coumadin (warfarin) for years. He had easy bruising (and may have bled into the wrist?). He had bladder stone removed 2–3 months earlier through cystoscopy. He also had exploratory laparotomy several years earlier, including lysis of adhesions. He had significant generalized arthritis and coronary artery disease (CAD), with coronary artery bypass graft (CABG) surgery more than 3 years earlier, plus cardiac stents 4 years previously. Other underlying diseases included hypertension, gout, and sleep apnea (he was on home continuous positive airway pressure – CPAP).

He was retired after working for many years in the construction business. He stopped smoking 40 years earlier, and stopped drinking alcohol 2 years before the office visit. He had no allergies.

On examination, he was alert, oriented, pleasant, and in no acute distress. The vital signs were as follows: blood pressure (BP) 148/78, respiratory rate (RR) 18, heart rate (HR) 78, temperature 98.6 °F; height 5′9″; weight 197 pounds.

Head and neck exam was significant for partial dentures with extensive dental work and bridges. No adenopathy was noted. The heart exam showed irregularly irregular heartbeat consistent with atrial fibrillation. The chest showed a healed thin median sternal scar from previous CABG surgery. The rest of the exam other than the extremities showed nothing significant. The upper extremities showed ecchymotic skin changes in the forearms; the right forearm, wrist, and proximal palm surgical site is shown in Fig. 1.2a. The radial artery was palpable.

- *What are the possible differential diagnoses based on the history and physical exam so far?*
- *What would be your recommendation to the orthopedic surgeon at this time?*

A preliminary result of the 2/16/09 wound culture was available, but final confirmation of the organism and sensitivity was still pending.

The ID consultant suggested a limited I&D of the ballotable bulge in the wrist, to resolve whether a hematoma, seroma or pus was present. Additional cultures would also be obtained.

Additional I&D was done the next day (3/5/09) and the patient was subsequently seen in the office 10 days after the initial office visit, 9 days after the second surgery.

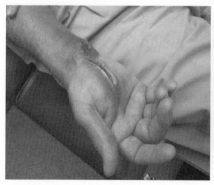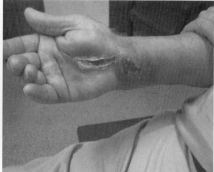

Figure 1.2a A large bulge of purplish bruised area is noted in the right wrist, along with the surgical incision in the hand. Photo was taken on 3/4/09 (reproduced with permission). (*See insert for color information*)

Initial laboratory findings

The histopathology of the 2/16/09 flexor tenosynovium from the right wrist was described as showing: "severe acute and chronic tenosynovitis with fibrin deposition, reactive and degenerative changes; rare giant cells with focal polarizable material and noncaseating granuloma."

• *How does this pathology report influence your preliminary diagnosis?*

Photographs of the right wrist were taken during follow-up visits 10 days, and 10 weeks, respectively, after the initial office visit on 3/4/09 (Fig. 1.2b).

The initial culture of tissue obtained on 2/16/09 was reported positive for acid-fast bacilli (AFB) on 2/26/09 (10 days later), and sent off for final identification or confirmation and antimicrobial sensitivity to a reference laboratory.

• *Final diagnosis: Mycobacterium marinum infection of right wrist*

The diagnosis was confirmed on 3/17/09, 1 month after the initial surgery and culture, by DNA probe and high performance liquid chromatography (HPLC).

The susceptibility test was available 18 days later, on 4/4/09. The organism was sensitive to all the six tested agents, including clarithromycin and trimethoprim/sulfamethoxazole, the two drugs which the patient was given as treatment. It was also sensitive to minocycline, doxycycline, ethambutol, and rifampin. Antimicrobial treatment duration was about 5 months, ending in August 2009.

Case discussion

This patient had *Mycobacterium marinum* infection in his right hand and wrist following a puncture injury from a shrimp fin that occurred when he was handling and cleaning shrimp. Initially, it was just a localized swelling of the thumb that was managed like a routine puncture wound. By the time he was seen by the ID

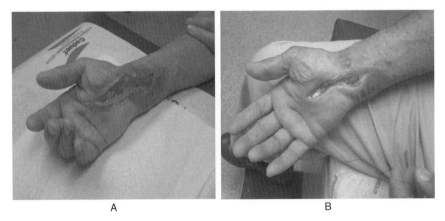

A   B

Figure 1.2b A. Photo taken on 3/14/09, 9 days after the second surgery. B. 10 weeks later, on 5/21/09 (reproduced with permission).

physician, he had been on clarithromycin for 3–4 weeks, and had been referred to the orthopedic surgeon who suspected a mycobacterial infection. However, the earlier treatment included, in addition to antibiotics, steroid injection in the wrist because of severe pain. The infection had spread to involve the hand and wrist, and included cellulitis of the forearm. All through these symptoms, fever was not an issue for the patient; pain and swelling were the paramount complaints. At incision and debridement (flexor tenosynovectomy) of the right wrist on 2/16/09, thick whitish pus was noted and cultured for routine as well as AFB and fungal cultures. All the initial stains were negative. However, 10 days later, an organism was cultured that was AFB positive. By the time of the office visit on 3/4/09, mycobacterial infection was the working diagnosis. The bulge in the wrist was opened up the next day, on 3/5/09, but that culture was negative. This patient was on warfarin, and may have also bled into the wrist area.

Mycobacterium marinum was suspected, based on the epidemiology (shrimp fin puncture injury, initial indolent infection, and AFB-positive growth in 10 days) of the wound and injury. Trimethoprim/sulfamethoxazole was therefore added on 3/6/09 to the clarithromycin that the patient was already taking. That choice was made based on knowledge of the local susceptibility pattern of this organism. As it turned out several weeks later, the M. marinum was sensitive to all six antimicrobial agents tested: clarithromycin, trimethoprim/sulfamethoxazole, minocycline, doxycycline, ethambutol, and rifampin.

The patient received the two-drug regimen of clarithromycin and trimethoprim/sulfamethoxazole from March to August, for 5 months altogether, and clarithromycin for slightly longer. The wound took so long to heal completely not because of bacterial persistence but because of the large size of the wound, two extensive surgical procedures, in a patient on warfarin therapy, and with other co-morbidities. He healed eventually without any plastic surgical procedures, but needed extended outpatient wound care 2–3 times a week. He was seen in the office at 3–6-weekly intervals, with the final visit 4 months after completion of antimicrobial therapy, and complete wound healing (Fig. 1.2c).

Non-tuberculous mycobacterial (NTM) infections

Currently, there are more than 120 species of mycobacteria classified under non-tuberculous mycobacteria (NTM) [1]. This group of mycobacteria is composed of species other than the M. tuberculosis complex, and had previous group names that included atypical mycobacteria and mycobacteria other than M. tuberculosis (MOTT) [1].

The classification of mycobacteria remains quite complex and continues to evolve, with advancement in culture and molecular techniques that have identified previously unknown organisms [1,2]. In the past, many of these organisms were either not recognized or were dismissed as laboratory contaminants. The

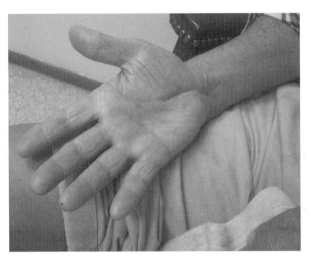

Figure 1.2c Healed right hand and wrist wound. Photo was taken on 12/18/09, 4 months after completion of antimicrobial therapy. The healing occurred without plastic surgery (reproduced with permission).

NTM are typically found in the environment, and have been recovered from surface water, including tap water and biofilms, soil, domestic and wild animals, as well as food and milk products [1,2].

Mycobacteria can be classified by their growth rate or whether they produce pigment [1,2]. The slowly growing mycobacteria require more than 7 days to reach maturity, while the intermediately growing NTM (e.g. *M. marinum* and *M. gordonae*) require 7–10 days to reach maturity. *M. marinum* grows optimally at room temperature, 28–30 °C [1,3].

Clinicians are, however, more familiar with the clinical syndromes produced by these mycobacteria. There are four distinct common clinical syndromes caused by NTM [1,2].
- Progressive pulmonary disease (bronchiectasis or chronic obstructive pulmonary disease [COPD]: *M. avium complex* [MAC], *M. kansasii*, and *M. abscessus*; especially in older adults).
- Superficial lymphadenitis (especially cervical in children caused by MAC, *M. scrofulaceum*, and *M. tuberculosis* in TB endemic areas).
- Disseminated disease (severely immunocompromised patients: HIV and non-HIV patients) [4].
- Skin and soft tissue infection (typically as a consequence of direct inoculation, as in our patient).

Mycobacterium marinum has in the past been recognized as causing soft tissue infections termed "swimming pool" and "fish tank" granuloma [1]. These designations reflect the type of activities those that acquired the infection were engaged in [5]. Skin break or inoculation precedes the infection, which typically follows within 2–3 weeks of contact or injury. The typical patient is otherwise healthy. Inoculation of the organism occurs through skin abrasion or punctures, and contact with saltwater fish, shrimp, or fins contaminated with *M. marinum*. The infection can sometimes present like sporotrichosis, with papules and nodules, but may produce more complicated wounds, as occurred in our patient. Antimicrobial treatment should take into account the local organism susceptibility, if available from the hospital or state health department laboratory.

Treatment typically includes two drugs, given for a minimum of 3 months [1]. Interestingly, an orthopedic group from Hong Kong reporting on *M. marinum* infection of the hand and wrist emphasized the importance of biopsy and debridement in managing patients with deep-seated infection [6]. In their review, they note that delay in diagnosis is common, because this is often an indolent infection. Often, the physician fails to elicit a history of aquatic exposure, and there is a lack of clinical suspicion, at least initially. A review of a large number of *M. marinum* infections was reported from France in 2002 [5]. Many of the reported cases had "fish tank" exposure, with most infections occurring in the upper limb.

Lessons learned from this case

- A good history and relevant epidemiology remain crucial in making a timely and accurate diagnosis of infectious diseases, including soft tissue infections such as mycobacterial infections.
- The injury from a marine crustacean (shrimp fin puncture) was a clue to the likely organism to expect in this wound infection.
- Appropriate cultures must be obtained, if an accurate diagnosis is to be made. If you do not think of the possibility, then the appropriate cultures (AFB) will not be done.
- Typically, appropriate mycobacterial cultures will require tissue obtained during surgical incision and debridement.
- Steroid injection, given for severe pain, was likely unhelpful and would be contraindicated in this case. I suspect it was done because the patient was not responding "fast enough" to the clarithromycin started earlier.
- Mycobacterial infections typically respond slowly, as with many indolent infections. Patience is therefore needed in the treatment of such infections.
- Often, a two-drug regimen is used, as in this instance, to prevent resistance during prolonged therapy.
- Treatment is typically prolonged: 3–4 months is usual, if there are no complications or underlying immunosuppression. With immunosuppression, treatment is even longer.

References

1 Brown-Elliott BA, Wallace Jr, RJ. Infections due to nontuberculous mycobacteria other than *Mycobacterium avium-intracellulare*. In: Mandell GL, Bennett JE, Dolin R (eds) *Principles and Practice of Infectious Diseases*, 7th edn. Philadelphia: Churchill Livingstone, 2010, pp. 3191–8.

2 Griffith GE, Wallace Jr, RJ. Pathogenesis of nontuberculous mycobacterial infections. Available at: www.uptodate.com/contents/pathogenesis-of-nontuberculous-mycobacterial-infections?topicKey=ID%2F5345&elapsedTimeMs=4&source=search_result&selectedTitle=9%7E150&view=print&displayedView=full, accessed February 24, 2016.

3 Griffith GE, Wallace Jr, RJ. Microbiology of nontuberculous mycobacteria. Available at: www.uptodate.com/contents/microbiology-of-nontuberculous-mycobacteria?topicKey=ID%2F5343&elapsedTimeMs=4&source=search_result&searchTerm=Overview+of+nontuberculous+mycobacteria&selectedTitle=15%7E150&view=print&displayedView=full, accessed February 24, 2016.

4 Griffith GE, Wallace Jr, RJ. Overview of nontuberculous mycobacterial infections in HIV-negative patients. Available at: www.uptodate.com/contents/overview-of-nontuberculous-mycobacterial-infections-in-hiv-negative-patients?topicKey=ID%2F5342&elapsedTimeMs=5&source=search_result&searchTerm=Overview+of+nontuberculous+mycobacteria&selectedTitle=1%7E150&view=print&displayedView=full, accessed February 24, 2016.

5 Aubry A, Chosidow O, Caumes E, Robert J, Cambau E. Sixty-three cases of mycobacterium marinum infection: clinical features, treatment, and antibiotic susceptibility of causative isolates. Archives of Internal Medicine 2002; 162(15): 1746–52. Available at: http://archinte.jamanetwork.com/article.aspx?articleid=754060, accessed February 24, 2016.

6 Cheung JP, Fung B, Samson Wong SS, Ip W. *Mycobacterium marinum* infection of the hand and wrist. Journal of Orthopaedic Surgery 2010; 18(1): 98–103. Available at: www.josonline.org/pdf/v18i1p98.pdf, accessed February 24, 2016.

The next four cases deal with severe beta-hemolytic streptococcal soft tissue infections.

Case 1.3   A 41-year-old Caucasian female with right arm swelling

A 41-year-old Caucasian female was admitted on 1/2/10, with acute swelling of the right arm and elbow. One day before admission she had complained of sore throat, followed by significant nausea and vomiting, up to 10 times in less than 24 hours. She awoke the next morning with redness and swelling of the right elbow. Within hours, the swelling had spread to involve areas below and above the elbow. A temperature of 103 °F was noted, in association with chills. She came to the emergency room, where she was medicated for nausea and vomiting, and admitted.

The review of systems was positive for sore throat starting the day before admission, fever and chills on the day of admission, and severe nausea and vomiting (10× in 24 hours) starting the day before and continuing up to the admission. On the day of admission, a rapid, severe, and extensive swelling of the right arm was noted.

The past medical history was significant for a previous swelling of the right elbow 2 years earlier without cellulitis. The symptoms resolved in 7 days. She had documented hypertension for several years and three C-sections 6, 10, and 15 years earlier. She did not drink alcohol, and had not smoked in 20 years. She was allergic to penicillin (rash).

Two days after admission, on January 4, infectious diseases consult was placed for assistance with the severe arm cellulitis, with bursa swelling of the right elbow.

The epidemiologic history was significant for the following: the patient's 10-year-old daughter had tonsillitis 7 days before the patient became ill with sore throat. By this time the daughter had recovered after antibiotic therapy. Her husband was also ill with a similar sore throat in the week between the daughter's illness and the patient's. Furthermore, she had cut her right index finger distal phalanx with a kitchen knife 8 days before her illness. Pus had developed a few days later, but the finger was now healed, with residual skin peeling but no streaking up the arm. They had two house dogs aged 3 and 9 years. She was a school teacher of 12–13-year-old children but school was out at Christmas, at the onset of the illnesses.

On examination, she was alert, oriented, not in acute distress, but worried. The vital signs were as follows: BP 95/52, RR 16, HR 70, temperature 100°F (up to 103.2°F the night before); height 5'6" and weight 160 pounds. The throat showed only slight redness of the oropharynx, but no pus. The heart, lung, and abdominal examinations were normal. Lymph nodes were not palpably enlarged. The right elbow was puffy and fluctuant, and there was tenderness involving the whole arm while the index finger was scabbed, reflecting the healing knife cut injury. Other extremities and the neurologic exam were normal. A photograph of the right arm taken on 1/6/10, 4 days after admission, is shown in Fig. 1.3a.

Laboratory findings showed the following complete blood count parameters and basic chemistries (sodium/potassium, glucose, BUN/creatinine, respectively) on the following days.

- On 1/2/10: white blood cells (WBC) 21.1, 12.3/36.2, mean corpuscular volume (MCV) 81.9, platelet 190 (diff: 89p 7L 4 M). Chemistries: 134/2.7, 125, 19/0.9
- On 1/3/10: WBC 18.4, 10.9/33.5, MCV 82.9, platelet 195 (83p 4b 6L 6M 1meta). Chemistries: 137/2.4, 104, 16/0.8
- On 1/4/10: WBC 14.4, 11.4/35.4, MCV 83.9, platelet 142 (55p 34b 7L 3M 1Eo). Chemistries: 136/4.1, 98, 14/0.8

Blood cultures on 1/2/10 and 1/4/10 were negative while plain x-ray of the right arm and elbow showed only soft tissue swelling.

Hospital course, procedures, and follow-up

A very limited incision and drainage procedure of the right elbow was done on 1/4/10. The photographs taken 2 days later on 1/6/10 (Fig. 1.3b) suggested a

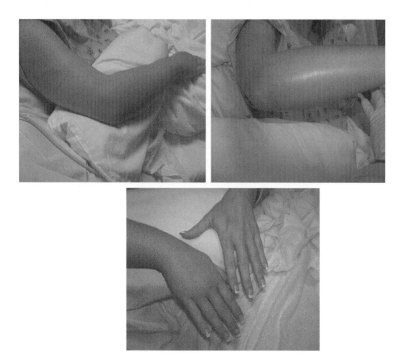

Figure 1.3a Swelling of the right arm, elbow, and hand, reflecting cellulitis. Compare the left hand without swelling. Photos taken on 1/6/10 (reproduced with permission). (*See insert for color information*)

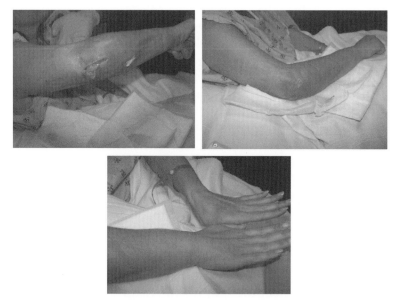

Figure 1.3b Photographs taken on 1/10/10, 2 days after the second surgery: markedly improved arm swelling (reproduced with permission).

need for further debridement. The patient was therefore taken back to surgery on 1/8/10 for a more extensive debridement. Two days later, there was a marked reduction in the arm swelling, reflected in the photographs taken on 10/1/10, 2 days after the second surgical debridement. The initial gram stain of pus obtained from the elbow on 1/4/10 showed 2+ (moderate) WBCs, but no organisms seen. Cultures from the elbow drainage on 1/4/10 and 1/7/10 were positive for group A beta-hemolytic *Streptococcus* (*S. pyogenes*).

Because of the history of penicillin allergy, the patient received a variety of other antibiotics that included daptomycin initially, and was discharged on 11/1/10 on a combination of oral clindamycin and levofloxacin. She continued to receive outpatient wound care till March 2010, 2 months after discharge from the hospital.

Comments

- This patient had inadequate surgical drainage initially, and so extended and prolonged inpatient and outpatient care continued till March 2010, 2 months after the hospital discharge.
- More detailed discussion will follow later after presentation of other cases.

Case 1.4   A 24-year-old female with fever and arm swelling

A 24-year old Caucasian female was admitted through the emergency room (ER) in late August 2009 because of swelling of the left elbow area.

Infectious diseases consult was sought 3 days later on 8/31/09 because of continued high fever, pain, and swelling of the arm.

The past medical history was significant for "crack cocaine" and other substance abuse, including oxycontin, and "crystal meth" (methamphetamine). She had failed a previous drug rehab treatment 2 years earlier. She had a history of anxiety and depression, asthma, hypertension, gastroesophageal reflux disease, and previous genital herpes infection. She had smoked a pack of cigarettes a day for 10 years and had no known drug allergies.

Epidemiologically important is that the patient had injected "crack cocaine" dissolved in water into her left arm (antecubital area) about 1 week before admission. Progressively worsening pain had been noted in the left elbow over the 3–4 days prior to admission.

On examination, she was alert, depressed, but not in acute distress. Her vital signs on 8/31/09 were as follows: BP 106/75, RR 16, HR 90, temperature 99.0 °F; height 5'8", and weight 136 pounds. The maximum temperature on 8/29/09, 1 day after admission, was 104 °F. Head and neck examination showed thrush while the heart, lung, and abdominal exams were unremarkable. She had tender left axillary adenopathy, but no other abnormal lymph node enlargements. The rest of the skin (other than the left arm) was healthy looking, except for tattoos. The neurologic exam was significant for a depressed affect, but no focal findings.

Hospital course, laboratory findings, and follow-up

The initial I&D of the left arm done on 8/28/09 was limited. The cultures obtained showed mixed organisms that included the following: group C *Streptococcus, Streptococcus mitis/oralis,* as well as diphtheroids. Additional I&Ds were done on 8/31/09, with obtained specimens showing a pure culture of group C *Streptococcus.* HIV serology was negative, as was serology for acute hepatitis A, B, and C. She was hospitalized from 8/28 to 9/5/09, a total of 8 days, before discharge to outpatient wound care and follow-up on oral amoxicillin/clavulanate and levofloxacin.

Comments

She was seen in the office 19 days after admission (9/16/09), with the wounds almost healed. She failed to show up for her scheduled final visit 2 months later.

I spoke to the patient by telephone nearly 6 years later in 2015. She was still struggling with aspects of drug addiction, but was well.

It is thought that she acquired her infection through injection drug use under unsterile conditions (crack cocaine or oxycontin dissolved in water). She required about three surgical debridement procedures before discharge from the hospital. Initial ultrasound examination of the arm on 8/28/09 suggested heterogeneous hypoechoic abscess or hematoma within the antecubitus. MRI 2 days later confirmed the presence of superficial perivascular and subcutaneous abscesses plus diffuse edema and enhancement consistent with cellulitis of the whole left arm.

Case 1.5   A 72-year-old male with necrotic soft tissue elbow infection

A 72-year-old black male was admitted with severe right arm and elbow cellulitis. He had been hospitalized for 9 days (1/26–2/5/10), and then transferred to a rehab center on 2/5/10.

Infectious diseases consult was placed on 2/16/10 to address the cause of the prolonged necrotizing fasciitis, unresolved after two previous radical debridement and fasciotomy procedures on 1/28 and 2/1/10.

Recent past history and epidemiology included the following: the patient had fallen at home, and was admitted and worked up for stroke and coronary artery disease during the admission of 1/2–1/4/10. He had three stents placed at that time. It was not clear how long he had lain on the floor after the fall. He was

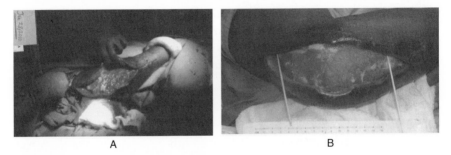

A   B

Figure 1.5a A. Photo taken at surgery on 1/28/10. B. Right elbow on 2/16/10, 19 days after the initial surgery (reproduced with permission).

readmitted from 1/7 to 1/18/10 with right arm cellulitis. Group A *Streptococcus* was cultured; clindamycin was prescribed, but it is uncertain how well the medication was taken by the patient. The third admission was on 1/26/10, this time for necrotizing fasciitis of the right arm and elbow.

Past medical history and other underlying diseases included type 2 diabetes mellitus, with cataracts and left eye blindness; degenerative joint disease (osteoarthritis); coronary artery disease with three stents placed 5–6 weeks earlier, and previous coronary artery bypass graft (CABG) surgery ×3. He had a recent stroke with left-sided weakness and paresis. He also had chronic kidney disease, with a creatinine value of 1.8 mg/dL.

On examination, he was alert and oriented. His vision was poor (fuzzy). His vital signs were as follows: BP 129/60, RR 18, HR 87, temperature 98.1 °F. His height was 5′6″ and weight 238 pounds. He had poor vision of the right eye with cataract, and loss of the left eye. He was edentulous. The heart, lung, and abdomen showed nothing significant while the chest showed a median sternal scar from previous CABG surgery. The upper extremities showed changes of osteoarthritis of the hands, loss of interosseous muscles of the left hand, and a slight weakness of the left arm, consistent with the stroke. The right arm showed a large open wound, up to 9 inches or more in length, as shown in Fig. 1.5a(B), taken on 2/16/10.

Comments

- The likely initial problem in the right arm was an abrasion suffered following a fall at home in early January, 2010. This may not have been readily apparent during the initial 1/2–1/4/10 admission. That hospitalization addressed primarily the stroke and CAD problems.
- Abscess was cultured on 1/7/10 when the patient was readmitted from 1/7 to 1/18/10 for cellulitis. However, he was inadequately treated until the admission on 1/26/10, during which period he had two I&Ds on 1/28 and 2/1/10.
- This patient was subsequently admitted to the hospital rehabilitation center from 2/5 to 2/25/10, and was discharged home on oral amoxicillin

and levofloxacin. He received hyperbaric oxygen therapy as part of the management of his arm wound infection during this admission.

- On 3/11/10, he was seen in the office with a wound VAC attached; the wound was still incompletely healed at that time, but was very clean. The antibiotics were discontinued 10 days later.

Case 1.6 A 37-year-old man with severe body aches and fever

A 37-year-old Caucasian male was admitted on 6/6/01 with severe headache, fever up to 100.5 °F, and confusion. CT scan of the head was normal. Lumbar puncture was attempted, but was unsuccessful, probably because of three previous back surgeries and distorted anatomy of the lumbar area. The WBC count was 12.1; the liver panel was normal, but throat culture was positive for group A *Streptococcus*. He was treated with levofloxacin and methylprednisolone, and discharged home on 6/8/2001, improved, on tapering steroids.

Two days after discharge home on 6/8/01, he developed new-onset fever and chills, generalized and severe body aches that became progressively worse, involving especially the right upper, right lower, and left lower extremities as well as the abdomen and back. He had no sore throat or cough at this time, and did not have a stiff neck or confusion. One day later, on 6/11/01, he was readmitted to the hospital.

The epidemiologic history showed that he had been covered with hundreds of gnats, especially on exposed legs, elbows and arms, after extended yard work 4 days prior to the initial admission on 6/6/01. Outpatient medications received by the patient were unknown.

On 6/12/01, one day after the second admission on 6/11/01, infectious diseases consult was sought for "septicemia."

The past medical history was significant for three back surgeries between 1984 and 1997, and hypertension. He was married, with a 15-month-old healthy son. He had a desk job, but spent a lot of time outdoors, including gardening. He had smoked one and a half packs of cigarettes per day for 20 years and was allergic to codeine (urinary retention).

On examination on 6/12/01, he was alert, apprehensive, and dyspneic. The vital signs were as follows: BP 136/66, RR 26, HR 99, temperature 102.3 °F; height 5'9" and weight 200 pounds. The head, neck, heart, back, lymph nodes, and genital examinations were unremarkable. Lungs showed left basal crepitations while the abdomen was distended, with active bowel sounds, but was non-tender. There was tense swelling of the right arm, from the biceps and triceps area to the elbow, with some blisters on the inner wrist and near the elbow. The right lower extremity showed tense swelling involving the thigh and calf, down to the foot. There were minimal changes in the left upper and left lower extremity.

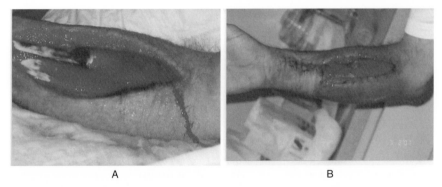

A   B

Figure 1.6a Right forearm: A. Day 15 post admission, after several debridements (note necrotic tendon). B. Day 21 of hospitalization: initial plastic surgery. Photos were taken on 6/26 (A) and 7/2/01 (B) (reproduced with permission). (*See insert for color information*)

   The skin showed signs of livedo reticularis, consistent with vasoconstriction or poor circulation (patient was, however, not on pressor agents at the time). Faint peripheral pulses were noted on the left foot. The neurologic exam was positive for brisk deep tendon reflexes in the lower extremities. He was generally weak, apprehensive, and able to move all extremities.

Comments and postscript

The patient was hospitalized from 6/11/01 to 7/2/01, and discharged to out-patient follow-up on day 21 following admission. He had at least four surgical debridement and fasciotomy procedures on the following days: 6/12, 6/14, 6/19, and 6/22/01 (Fig. 1.6a). Additional skin grafting/plastic surgery would be done in the outpatient setting later, to ensure closure of the multiple wounds in all four extremities. While hospitalized, he received various antimicrobial agents that included, initially, piperacillin/tazobactam and clindamycin. He was subse-quently discharged home on oral amoxicillin and ciprofloxacin. He had addi-tional plastic surgery procedures in the outpatient setting and was completely healed when seen in the office 10 weeks later (Fig. 1.6b). A chest x-ray obtained on 8/23/01 showed complete clearance of the left-sided pneumonia and effu-sion. By the time of that office visit, he had been back to work (desk job) for several weeks.

   The characteristics of the four patients with severe beta-hemolytic *Streptococcus* infection are shown in comparative detail in Table 1.6a.

Case discussion of the four beta-hemolytic *Streptococcus* infections

All four patients with beta-hemolytic *Streptococcus* infections presented here had severe illnesses that required either prolonged hospitalization or prolonged

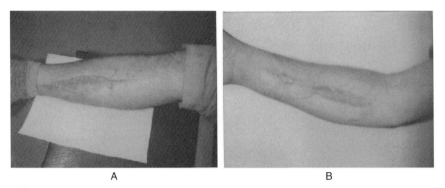

A   B

Figure 1.6b Photos taken on 8/23/01. Office visit 10 weeks after hospital admission. A. Healed right leg. B. Healed right arm wounds following skin grafts (reproduced with permission).

outpatient follow-up for wound care. They all also had multiple surgical debridement (at least 3× in each case) before the infection could be brought under control. Three cases were due to group A *Streptococcus*, while one case (case #1.4) was due to group C *Streptococcus* infection. The first and third cases were due to group A *Streptococcus* (*S. pyogenes*); the second case was due to group C *Streptococcus*. All of these streptococci are beta-hemolytic. The underlying pathogenesis in each case was some form of trauma: cut, abrasion or penetrating injury, or injection (as in the case of patient #1.4, who was an IV drug abuser). Patient no. 3 (case #1.5) had fallen after a stroke and likely lay on the floor for an extended period of time at home with a skin abrasion. His skin trauma or injury was missed, or inadequately cared for during the initial admission, because there was more concern at the time for the stroke and possible vascular event that may have led to the stroke. He had three coronary stents placed during the January 2–4, 2010 admission. He was readmitted 3 days later with right arm cellulitis, at which time *S. pyogenes* was cultured. He then had the first of three surgical incision and drainage or fasciotomy procedures in that arm. The third admission on 1/26/10 was for necrotizing fasciitis of that same arm.

The fourth patient (case #1.6) had group A streptococcal myonecrosis (gangrenous myositis), a different and more devastating clinical syndrome. This patient had multiple, extensive surgical debridement or fasciotomies during the second admission to the hospital and a prolonged second hospital stay of 21 days. He had severe body aches and systemic symptoms. Group A *Streptococcus* was cultured from the throat, blood, and wound (arm and leg). His was a more virulent disease with a higher mortality than the preceding three cases. All of the four presented patients survived.

The characteristics of the four cases of severe beta-hemolytic streptococcal soft tissue infections presented here have been summarized in Table 1.6a. All of these cases (except case #1.5 which occurred following a stroke) presented acutely

Table 1.6a Characterization of patients with severe beta-hemolytic streptococcal soft tissue infections.

Characteristic	Case #1.3 Admitted 1/2/10	Case #1.4 Admitted 8/28/09	Case #1.5 Admitted × 3: 1/2, 1/7, and 1/26/10	Case #1.6 Admitted × 2: 6/6 and 6/11/01
Age/sex	41 year C/F	24 year C/F	72 year B/M	37 year C/M
Site of infection	Right elbow	Left elbow	Right elbow	Right arm, right leg, and left leg
Maximum temperature (°F)	103.2	104	98.9	102.3
Days of symptoms before admission	1–2 days	3–4 days	1 day before initial admission; 3 days before second admission	4 days before 6/6/01 admission; 1 day before readmission on 6/11/01
Symptoms experienced	Sore throat, pain + swelling right elbow, N/V, and fever	Severe pain left elbow, and fever	Abrasion, pain, abscess, and swelling R elbow	Severe body aches, fever, and confusion
Duration of hospitalization	9 days	8 days	11 days	21 days
Number of surgical procedures and type	2×: incision and debridement	3×: incision and debridement	3×: incision and debridement, and fasciotomy	>4×: incision and debridement, or fasciotomy
Duration of outpatient follow-up	>2 months	19 days	>2 months	>2 months
Epidemiology of infection	Daughter and husband with sore throat; cut right index finger 10 days PTA	IV drug abuse: crack cocaine injection	Abrasion right arm following a fall; inadequate initial I&D, and treatment	Hundreds of gnats on exposed skin after extended yard work; presumably skin abrasions
Underlying diseases	Hypertension, allergy to penicillin	Substance abuse, GERD, and hypertension	DM type II, CAD, CABG surgery, CVA, CKD, and DJD	Hypertension, 3 back surgeries, and smoker

(continued)

Table 1.6a (*Continued*)

Characteristic	Case #1.3 Admitted 1/2/10	Case #1.4 Admitted 8/28/09	Case #1.5 Admitted × 3: 1/2, 1/7, and 1/26/10	Case #1.6 Admitted × 2: 6/6 and 6/11/01
Organism isolated	Group A *Streptococcus*	Group C *Streptococcus*	Group A *Streptococcus*	Group A *Streptococcus*
Other laboratory findings	WBC 21.0; blood cultures negative	WBC 12.9; blood cultures negative	WBC 9.7; blood culture not done	Blood cultures × 3 + for GAS; wound and throat cultures + for GAS; CPK 3166 (elevated)
Clinical syndrome	Severe cellulitis	Necrotizing fasciitis	Necrotizing fasciitis	Gangrenous myositis (myonecrosis)
Complications	Prolonged outpatient wound care	Extensive I&D above and below left elbow	3 hospital admissions, and 3 I&Ds	>4 extensive I&Ds and fasciotomy; left lower lung pneumonia and effusion; skin graftings
Antimicrobial therapy	Daptomycin, clindamycin, and levofloxacin	Ceftriaxone, amoxicillin/clavulanic acid, and levofloxacin	Vancomycin, amoxicillin, clindamycin, and levofloxacin	Piperacillin/tazobactam, clindamycin; amoxicillin and ciprofloxacin
Outcome: survived/died	Survived	Survived	Survived	Survived

B, black; C, Caucasian; CABG, coronary artery bypass graft; CAD, coronary artery disease; CKD, chronic kidney disease; CPK, creatine phosphokinase; CVA, cerebrovascular accident; DID, degenerative joint disease; DM, diabetes mellitus; GAS, group A *Streptococcus*; GERD, gastroesophageal reflux disease; I&D, incision and debridement; M/F, male/female; N/V, nausea/vomiting; PTA, prior to admission; WBC, white blood cell count.

(within days) with systemic symptoms that included high fever, body aches, and pain. All patients required multiple surgical debridement or fasciotomy procedures and prolonged inpatient or outpatient follow-up wound care management, in order to accomplish wound healing. The multiple antimicrobial agents used were a reflection of the severity of the patients' illnesses, and the desire on the part of the clinicians to find the "best" regimen.

Most beta-hemolytic streptococci are sensitive to most of the antimicrobials available, other than macrolides, e.g. erythromycin [1]. The isolates in all of our four patients were sensitive to the agents used for treatment. This group of organisms has maintained a relatively predictable susceptibility. The major toxicities caused by the beta-hemolytic streptococci are typically due to toxin production and other virulent factors, and not to any high level of antimicrobial resistance. Stevens and Bryant have outlined some of the virulent factors and other pathogenic mechanisms of disease caused by group A *Streptococcus* [2].

Adequate surgical debridement appears to be very important in achieving a more rapid healing, while inadequate debridement often led to prolonged hospitalization and wound healing.

In humans, the two most common sites of group A *Streptococcus* (GAS) infection are the skin and the upper respiratory tract. The intact skin usually presents an effective barrier to colonization or infection with GAS. However, infection is facilitated when there is skin damage [3], as was the case in all of the four patients described above. Because drainage of the abscesses and necrotic tissue was initially inadequate in all four patients, the duration of illness was prolonged, requiring additional surgical procedures. The patient with gangrenous myositis (myonecrosis) (case #1.6) had a more virulent disease, typically associated with a higher mortality. He was the only patient who had a positive blood culture, in addition to positive GAS cultures in wound tissue, and earlier in the throat as well. The differential diagnoses of GAS infections involving muscle and fascia have been summarized by Stevens and Baddour [4].

Lessons learned from these cases

- The typical severe beta-hemolytic streptococcal soft tissue infection presents acutely (within days), with systemic symptoms of fever, body aches, pain, and swelling.
- The source of most soft tissue infections is typically a breach in the skin by a variety of mechanisms, as was the case in all four patients presented here.
- Adequate drainage of skin abscess or necrosis is a crucial part of the management of severe beta-hemolytic streptococcal soft tissue skin infections.
- Delay in incision and drainage or incomplete drainage will likely lead to prolonged hospitalization or extended outpatient therapy.

- Antimicrobial therapy alone is often not sufficient to achieve successful treatment of severe soft tissue infections.
- The injury or damage caused by this group of organisms is usually not due to their high level of antimicrobial resistance, but to virulent factors that they produce, host factors, and type of treatment available to a patient.

References

1 Stevens DL, Bisno AL, Chambers HF, et al. IDSA practice guidelines for the diagnosis and management of skin and soft-tissue infections. Clinical Infectious Diseases 2005; 41: 1373–406.

2 Available at: www.idsociety.org/uploadedFiles/IDSA/Guidelines-Patient_Care/PDF_Library/ Skin%20and%20Soft%20Tissue.pdf, accessed February 24, 2016.

3 Stevens DL, Bryant A. Group A streptococcus: virulent factors and pathogenic mechanisms. Available at: www.uptodate.com/contents/group-a-streptococcus-virulence-factors-and-pathogenic-mechanisms?view=print, accessed February 24, 2016.

4 Leyden JJ, Stewart R, Kligman AM. Experimental infections with group A Streptococci in humans. Journal of Investigative Dermatology 1980; 75: 196–201.

5 Stevens DL, Baddour LM. Necrotizing soft tissue infections. Available at: www.uptodate.com/ contents/necrotizing-soft-tissue-infections?topicKey=ID%2F7662&elapsedTimeMs=5& source=search_result&searchTerm=group+a+beta+hemolytic+strep&selectedTitle=11 %7E150&view=print&displayedView=full, accessed February 24, 2016.

Further reading

Bisno AL, Stevens DL. Streptococcal infections of skin and soft tissues. New England Journal of Medicine 1996; 334(4): 240–5.

Bisno AL, Stevens DL. Streptococcus pyogenes. In: Mandell GL, Bennett JE, Dolin R (eds) *Principles and Practice of Infectious Diseases,* 7th edn. Philadelphia: Churchill Livingstone, 2010, pp. 2593–610.

Carey RB. Identification of the crazy cocci: how and when. Available at: www.swacm.org /annualmeeting/2011/handouts/carey_01.pdf, accessed February 24, 2016.

CHAPTER 2

Fever of Unknown Origin and Drug-Induced Fever

Case 2.1   Hypersensitivity masquerading as sepsis

A 59-year-old black female was brought to the emergency room (ER) in May, 2005 because of weakness, shortness of breath, confusion, and fever. She was a poor historian because of a previous head injury 30 years earlier. While still in the ER, she became hypotensive, and with a temperature of 104.1 °F, she was thus admitted to the medical intensive care unit (ICU).

Her past medical history was significant for a recent 4-day admission for suspected sepsis, with hypotension and fever. She had been discharged 1 day before on oral antimicrobials to complete treatment at home. Her blood cultures had been negative, but a left middle toe culture was reported positive for MRSA a few days before that admission. Two months earlier, she had been admitted for left flank subcutaneous abscess with MRSA. That abscess had been drained, and was now healed.

At age 20, she had a motor vehicle accident, resulting in head injury, surgery for intracranial bleed, and residual right hemiparesis and expressive dysphasia and dysarthria. Other illnesses included a controlled seizure disorder secondary to the head injury, hypertension, and gastroesophageal reflux disease. Surgical procedures included hysterectomy, cholecystectomy, and a left knee surgery. She had a negative persantin cardiolite study 3 years earlier.

The family and social history was significant for no alcoholism. She did not smoke, but dipped snuff. Although she did live on her own prior to the onset of her illness, she had been staying with her sister since her recent illnesses with recurrent admissions. She was unmarried. Her sister had power of attorney in major decisions because of her head injury-related deficits. She had no known drug allergies.

Medications prior to admission included aspirin, trimethoprim/sulfamethoxazole, rifampin, phenobarbital, and PRN metoclopramide.

Three days after admission, infectious disease consult was sought for "recurrent sepsis."

Cases in Clinical Infectious Disease Practice: Obtaining a Good History from the Patient Remains the Cornerstone of an Accurate Clinical Diagnosis: Lessons Learned in Many Years of Clinical Practice, First Edition. Okechukwu Ekenna.
© 2016 John Wiley & Sons, Inc. Published 2016 by John Wiley & Sons, Inc.

On examination, she was alert, oriented to person, and not in acute distress. Her vital signs showed the following: blood pressure (BP) 102/64, respiratory rate (RR) 26, heart rate (HR) 96, temperature 99.5 °F; height 5′6″, weight 220 pounds.

The head and neck, heart, and lung exams were unremarkable. The abdomen was obese, and showed a healed right upper quadrant scar from previous cholecystectomy. The right upper extremity showed flexion contracture at the elbow. The lower extremities showed osteoarthritic changes at the knees. Lymph nodes were not enlarged and the skin showed no rash. Neurologic exam showed dysarthric speech and fluctuant mood and affect (friendly, sad, suspicious, and co-operative at various times). She was, however, able to follow simple commands. Deep tendon reflexes were increased in the right lower extremity, consistent with the previous stroke equivalent (head injury with intracranial bleed).

Laboratory parameters included the following. On admission through the ER the white blood cell count (WBC) was 3.0/μL, hemoglobin and hematocrit (H/H) 12/38, respectively, mean corpuscular volume (MCV) 91, and platelet count 357,000/μL. Differential count showed 69% polymorphs, 27% lymphocytes, and 2% monocytes. Total creatine phosphokinase (CPK) was 568 (mildly elevated), with borderline creatine kinase myocardial band (CK-MB) and troponin I levels. Blood cultures were negative during this and four previous admissions in May, 2005, and previously in March, 2005. *Staphylococcus aureus* (MRSA) was cultured from the left flank abscess in March and from the left middle toe in May, several days prior to this current admission. Erythrocyte sedimentation rate (ESR) was 35 mm/hour, C-reactive protein was 11.79 (N<0.3 mg/dL), liver and renal function tests showed only mild changes, and basic natriuretic protein (BNP) was 94 (normal).

Chest x-ray was clear, and computed tomography (CT) scans of the chest, abdomen, and pelvis were negative. Bone scan showed no osteomyelitis while CT scan of the head confirmed an old left middle cerebral artery (MCA) infarct.

Hospital course

After extensive chart review and patient evaluation, the infectious diseases (ID) consultant made the following observations, and compiled a list of signs and symptoms exhibited by the patient just before and during recent hospital admissions as follows.

- Disorientation, confusion, weakness
- Shortness of breath (SOB)/wheezing
- Hypotension, systolic BP as low as 69 mmHg
- Rapid heart rate, up to 155
- Rapid respiratory rate
- Fever, up to maximum temperature 104 °F

- Rash and itching

He also reviewed the drug treatment history, and noted that the patient was usually improved by the time she was discharged to outpatient follow-up to complete treatment for "sepsis" on oral drugs. He thought there was a good case here for drug-induced hypersensitivity, and not a true sepsis. Trimethoprim/sulfamethoxazole was thought to be the most likely drug causing these symptoms.

To confirm this would require challenging the patient with the suspected drug or agent under controlled conditions. This issue was discussed with the patient's sister (who had power of attorney for the patient), the attending physician, a cardiologist, and the nursing staff caring for the patient.

Hypothesis testing by drug challenge
Preparation for drug challenge

- Signed informed consent was obtained (from the sister, who had power of attorney).
- Notification of ICU nurse, attending physician, and cardiologist.
- Patient situated in the MICU.
- Establishment of secure central venous access.
- Provision of diphenhydramine, methylprednisolone, and normal saline for infusion.
- Trimethoprim/sulfamethoxazole tablet and suspension.

Other conditions to be fulfilled during the drug challenge

- Personal presence/availability for several hours for patient observation and intervention if needed (infectious diseases consultant overseeing the drug challenge).
- Nasal oxygen at 2 L all the time.
- Presence of the sister or other family member during most of the period of challenge and observation for reassurance of the patient.
- Communication with the team (nursing, attending, cardiologist, family) throughout the period of challenge/observation.

Results of drug challenge (recorded observations and interventions) (Figs 2.1a–2.1c)

The next day, the patient was given a trimethoprim/sulfamethoxazole (TMP-SMX) tablet (preferred by patient over suspension) at 08.25 am, with the ID consultant and the patient's sister present together in the medical ICU. She then proceeded to exhibit all the signs and symptoms recorded in the chart below. The lowest BP was noted at 73/42, highest respiratory rate was 50/minute, and the maximum temperature was 103.8 °F. She also developed erythematous skin flush with itching and scratching, and had tremors and

DATE	June 1, 2005					
0825	BP 114/94	Pulse 93	Resp. 17	Temp 98.5	Bactrim 85 po	100%
0840	BP 104/66	96	20	98.7	Tremors/vitals	100%
0855	116/76	98	21	98.9	Trem/shakes disorientation	100%
0905	105/83	107	20	99.8	" "	100%
0930	149/83	111	22	99.3	" flexing	100%
0945	147/82	113	28	99.1	"	100%
1000	132/76	129	40	99.1	" resp/airway	100%
1000	Give Salumedral 125 mg IV Benadryl 25 mg IV					
1015	145/77	147	48	99.9	tremors/skin	98%
1030	73/52	155	60	100.3	resp labored "	90%
1030	Loppressor 5 mg IV given NS 999/hr					
1035	92/60	128			eyes red watering	
1045	105/59	141		102.0	tremors/resp labored	88%
	Tylenol supp. given					
1100	92/60	148	48	102.7	resp labored	100%
1115	109/70	148	48	103.1	red eyes/tremor resp + labored	100%
1130	105/61	145	50	103.6	less tremors	100%
1145	119/87	141	46	105.8	resp ↓ labored	100%
1200	78/48	134	36	103.4	" "	100%
1145	NS + 500 cc/hr BP ↓ 78/48					
1200	NS ↑ to 999 cc/hr Dr Ekenna notified					
1225	80/45	133	36		no tremors resp rapid non labored	100%
1240	83/41	130	39			
1256	97/46	129	28			
1311	97/51	127	30	102.4		
1341	94/45	127	23			
1425	85/48	129	26		Red rash generalized itching	
1430	neosynephrine drip resumed NS 999 D/ced				Benadryl 1500 Salumedral 125 1600	

Figure 2.1a Vital signs and other comments recorded on the day of drug challenge, 6/1/05.

disorientation within the first hour of taking the TMP-SMX. The heart rate reached a peak of 155 beats per minute.

Interventions included the use of large volumes of normal saline (with low BP), methylprednisolone and diphenhydramine (with skin flush, rash, itching, and dyspnea/wheezing), and beta-blocker (for severe tachycardia).

By 6 hours most of the parameters were back close to base line, and by the next morning, all the adverse symptoms had nearly completely resolved.

Summary of signs and symptoms experienced during drug challenge with TMP-SMX

- Disorientation, restlessness, tremor, confusion, weakness
- Tachycardia up to 160 beats/minute
- Increased respiratory rate up to 50/minute, with wheezing
- Hypotension, systolic of 73 mmHg
- Fever, maximum temperature of 103.8 °F
- Diffuse erythematous rash (flush)
- Severe pruritus
- Conjunctival injection or redness

The signs and symptoms observed during the drug challenge were completely similar to the ones exhibited by the patient just prior to admission (while on TMP-SMX therapy) and during the first day(s) of admission, before the

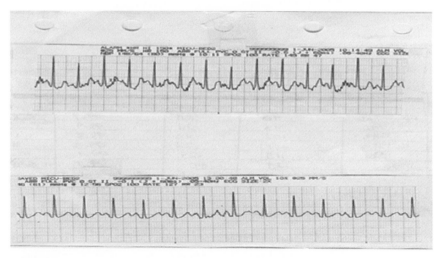

Figure 2.1b Sample of EKG tracing noted during drug challenge on 6/1/05 (the maximum rate noted on the top tracing was about 150 beats per minute).

oral agents were changed to parenteral treatment with vancomycin and piperacillin/tazobactam. Even the sequence of symptom onset was similar.

Twenty-four hours after the drug challenge, all the antimicrobial therapies were discontinued. There was no recurrence of the fever or the other symptoms. The patient was given an allergy armband for "SULFA" (trimethoprim/sulfamethoxazole), with instructions to the family that she was allergic to this medication and should not receive this drug in the future.

She was discharged home several days later.

• *Final diagnosis: Hypersensitivity to trimethoprim/sulfamethoxazole, masquerading as sepsis*

Case discussion

This patient was previously admitted in March, 2005 with a left flank cutaneous abscess that was positive for *Staphylococcus aureus*. The abscess was drained and treated with vancomycin in the hospital, and she was then discharged home on oral levofloxacin and rifampin. She was considered cured from that infection.

She was again seen as an outpatient in May, 2005 for a left middle toe abscess, and placed empirically on a combination of oral levofloxacin and trimethoprim/sulfamethoxazole after cultures were obtained. Two days later she returned to the ER with symptoms of "sepsis" (low BP, confusion, fever up to 104°F, plus other symptoms). That culture was subsequently positive for MRSA. After parenteral therapy in the hospital that included vancomycin and levofloxacin, she improved and was subsequently discharged home on a combination of trimethoprim/sulfamethoxazole and rifampin after 4 days.

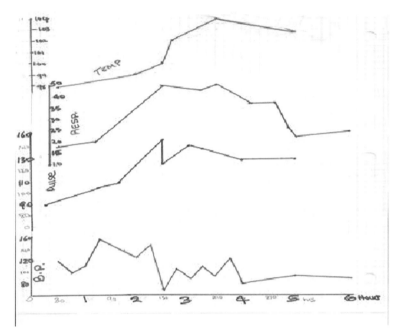

Figure 2.1c Graphic display of temperature, respiration rate, pulse rate, and blood pressure during drug challenge with trimethoprim/sulfamethoxazole over a period of 6 hours.

Her readmission this time was only 1 day after the most recent discharge in late May, 2005. That admission, again with symptoms of "sepsis," is the one discussed above.

A careful review of the record of events, the medications, the sequence/time relationship of the symptoms to the treatments received as inpatient and outpatient led us to formulate the hypothesis and the testing described above.

Additional finding in the laboratory parameters showed that 1 day following admission to the ER, when receiving steroid therapy to address the symptoms of wheezing, erythematous skin flush, etc., the WBC count rose from normal to 35,000/μL. CPK rose from a mild elevation of 568 to 3243. The same pattern of elevation of WBC and CPK was repeated 1 day after the drug challenge (24,700 and 2162, respectively, for WBC and CPK). The clinical parameters thus paralleled the laboratory findings in both instances. By day 7 post challenge, all the laboratory parameters were back to baseline.

We think that Koch's postulate was met in this case, and that a hypothesis was formulated, properly tested, and proved correct.

Figure 2.1d is an illustration of some of the mechanisms involved in drug-induced fever.

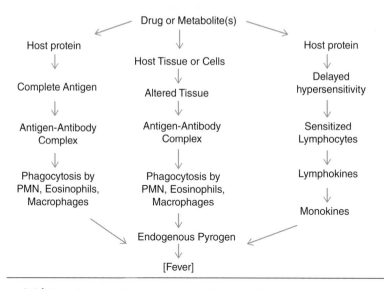

Figure 2.1d Three basic mechanisms postulated for the explanation of drug-induced fever in humans. PMN, polymorphonuclear leukocyte. Reproduced with permission of Oxford University Press from Young et al. [1].

Lessons learned from this case

- This case illustrates the importance of careful and detailed history taking.
- Drug-induced hypersensitivity can present exactly like sepsis. The pathophysiology is quite similar.
- The review of the patient's recent admission records was the key to formulating a hypothesis.
- Testing such a hypothesis required the presence of the physician (in this case, the ID consultant) to both observe the symptoms and to take appropriate life-saving action if necessary. A team approach was needed, and worked well here.
- The oversight of this procedure should not be delegated to a junior staff member or a non-physician. The cardiologist was available, in case critical care help was needed.
- If a decision is made to test out such a hypothesis, then the physician must take the time necessary to do the job well.
- Although this was stressful to the patient, it was important to establish the cause of the recurrent admissions, and thus avoid unnecessary treatment with potentially dangerous medications, as well as the cost of multiple avoidable admissions.
- Needless to say, full informed consent must be obtained, and adequate preparation made before such a procedure is embarked upon.

Reference

1 Young EJ, Fainstein V, Musher DM. Drug-induced fever: cases seen in the evaluation of unexplained fever in a general hospital population. Review of Infectious Diseases 1982; 4: 69–77.

Further reading

Bone RC, Balk RA, Cerra FB, et al. Definitions for sepsis and organ failure and guidelines for the use of innovative therapies in sepsis. Chest 1992; 101:1644–55.

Mackowiak PA. Concepts of fever. Archives of Internal Medicine 1998; 158: 1870–81.

Munford RS, Suffredini AF. Sepsis, severe sepsis, and septic shock. In: Mandell GL, Bennett JE, Dolin R (eds) *Principles and Practice of Infectious Diseases*, 7th edn. Philadelphia: Churchill Livingstone, 2010, pp. 987–1010.

Neviere R. Sepsis and the inflammatory response syndrome: definitions, epidemiology and prognosis. Available at: www.uptodate.com/online/content/topic.do?topicKey=cc_medi/9693&view=print, accessed February 24, 2016.

Russell JA. Management of sepsis. New England Journal of Medicine 2006; 355: 1699–713.

Case 2.2   A psychiatric patient with HIV infection and fever

A 36-year-old black male was admitted to the psychiatry service in April 1998 because of acute psychosis. He complained of having auditory and visual hallucinations, and felt like "the spirit is in me." He was brought to the hospital for help with his psychosis.

He was diagnosed with HIV infection in 1992 while living in Los Angeles, California. He was treated with zidovudine for 3–4 months and stopped taking the drug because of both cost and tolerance issues (side effects). He went for many years without specific HIV treatment.

He was restarted on a new "cocktail" of HIV medications 1 month prior to admission. When his medications started in March, 1998, his HIV-1 viral load was 122,015 copies/mL, and CD4 count was only 3. The combination drugs included zidovudine, lamivudine, and saquinavir (Fortovase).

His past medical history included the following: HIV infection diagnosed in 1992, hospitalization for pneumonia in 1995, rectal surgery for warts in 1997, and long-standing, mild non-specific dermatitis for years.

He was single and homosexual. Up until March, 1998, he was actively working as an airline attendant, but took sick leave once the anti-HIV medications were started. He did not smoke, drink alcohol, or use recreational drugs. There was a history of questionable allergy to "sulfa."

His medications prior to admission included azithromycin, zidovudine, lamivudine, and saquinavir, as well as trimethoprim/sulfamethoxazole (TMP-SMX), and clotrimazole troche.

The review of systems was significant for no fever, but some unspecified weight loss, general asthenia, and the recent auditory and visual hallucinations.

Infectious diseases consult was placed 1 day after admission for management of the HIV infection.

On examination, he was noted to be asthenic, and in no acute distress. His vital signs were as follows: BP 124/76, RR 24, HR 77, temperature 98.4 °F; height 5'11", weight 121.5 pounds. His oxygen saturation at room air was 96%.

Head and neck exam showed pharyngeal thrush. Heart, lung, and abdominal exams were normal. The skin showed dark spots of old chronic acneiform dermatitis but no new rash, and lymph nodes were palpably enlarged in the inguinal and epitrochlear areas. Neurologic exam found him to be nervous, but otherwise there were no focal findings.

Hospital course

The patient was seen by the psychiatrist and the psychosis was managed with various medications, including risperidone. Other in-hospital medications included

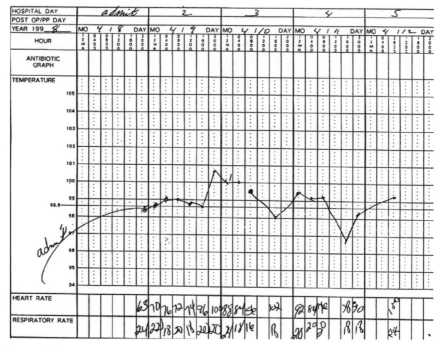

Figure 2.2a Temperature recordings from 4/8 to 4/12/98, with heart and respiratory rate notations.

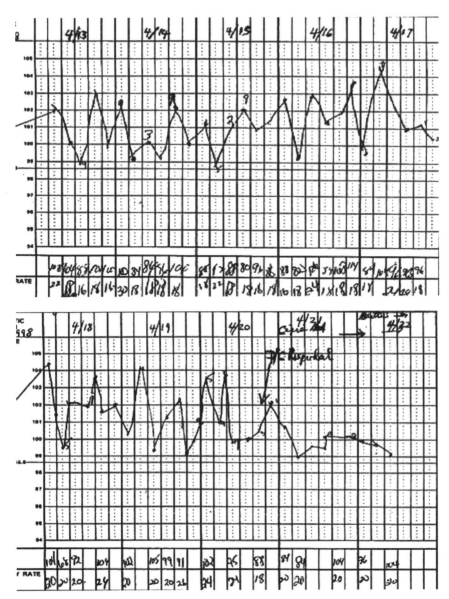

Figure 2.2b Temperature recordings from 4/13 to 4/22/98, with heart and respiratory rate notations.

the home medications noted above. The infectious diseases consultant was asked to review the HIV medications for any adjustments. The recording of temperature and other vital signs in the first 4 days is shown in Fig. 2.2a.

The temperature was either normal or low grade in the first 4 days of admission. However, beginning on day 5 of admission, and for the next 8 days

(4/13 to 4/20/98), there was a persistent fever, with spikes between 102 °F to 104.7 °F (Fig. 2.2b).

As part of the work-up of the persistent fever, extensive investigations were done. A test for HIV-1 viral load found that he had undetectable virus (<25 copies/mL), while the CD4 count had gone up to 20 from 3. The following studies were negative or normal: multiple chest and sinus x-rays, CT scan and MRI of the head, blood cultures ×8, urinalysis and culture, cytomegalovirus (CMV) and Epstein–Barr virus (EBV) serology, as well as sputum cultures. Echocardiogram was normal, and negative for vegetations. A lumbar puncture done because of concern for the psychosis was negative (zero cells, negative culture). Chemistries were normal, including liver tests. However, complete blood count (CBC) showed leukopenia, with WBC in the range of 1.7–3.7, H/H 9.3/27.9, respectively, and platelet count 405,000. The differentials showed 70% polymorphs, 1% band form, 14% lymphocytes, and 14% monocytes.

Antimicrobial therapy did not control the fever. There was concern that the fever could be related to one of the new HIV medications that had been started only 3–4 weeks earlier, or perhaps the TMP-SMX (questionable allergy history). These and other medications were stopped, but the fever continued.

Finally, on 4/21/98, on day 13 of admission (8 straight days of high fever), a decision was made to stop one of the antipsychotic medications (risperidone). The fever resolved in the next 24 hours after that.

- *What is your diagnosis?*

The patient was discharged home several days later with resolved fever, on the following medications: the antiretroviral drugs, loratadine, nystatin, azithromycin, and trimethoprim/sulfamethoxazole. Quetiapine was to be used only in case of hallucination.

- *Final diagnosis: Drug-induced fever secondary to risperidone.*

Case discussion

This case illustrates several important points relating to drug-induced fever in HIV-infected patients. First, HIV infection is a cause of fever. Infected patients also have a propensity to experience adverse reactions, including fever, to many medications. This patient had been started on a combination of three antiretroviral drugs 3 weeks before admission, any one of which could be a culprit. His history suggested a possible drug reaction to TMP-SMX in the past, a drug he was placed on for *P. jiroveci* (PCP) prophylaxis. He also came into the hospital because of psychosis, and so there was a great reluctance on the part of the psychiatrist, at least initially, to consider stopping one of the antipsychotic drugs when this possibility was broached. It was only after we had tried stopping various other medications, including the antiretroviral drugs, that the psychiatrist agreed to stop the risperidone. The patient defervesced within 24 hours after the drug was stopped. We were able to resume all the other necessary medications, including the antiretroviral drugs and TMP-SMX, without recurrence of the fever.

The pattern of the fever shows that temperature level was not a useful differential, as this patient had temperature spikes between 102°F and 104.7°F. However, the rest of the vital signs remained stable, with respiratory rate normal and pulse rate inappropriately normal for the high temperatures recorded. That was another subtle clue that drug fever was a possibility.

Drug-induced fever can be expensive to investigate, especially in a patient with immunosuppression, as in this case. Multiple scans, x-rays, and cultures were needed to rule out rare or more subtle infections, or unusual presentation of a common infection or disease.

Although good clinical skills and observation are very helpful, in the final analysis, drug-induced fever is often a diagnosis of exclusion.

More comparative analyses and discussion of drug fever will follow after presentation of the next case.

Case 2.3   A hypertensive patient with 2 months of fever and chills

A 49-year-old black male was admitted in late April, 1998 with 2 months of fever, chills, and sweats. He had complained of generalized malaise, fever, and night sweats, requiring change of clothing several times during the day and night. He also had chills and rigors. In between the fever and chills he felt generally well. These symptoms had been going on for about 2 months.

Several weeks before admission, he had esophagogastroduodenoscopy (EGD) and colonoscopy performed because of blood in the stool. Gastroesophageal reflux disease (GERD) was noted, and he was placed on omeprazole. His past medical history included hypertension that had been diagnosed for 15 years. He had complained of chronic sinusitis with stuffiness of the nose, but was otherwise healthy.

He was married and worked in an office setting. Although there were five grown children living away from home, the family had adopted a 28-month-old girl with chronic respiratory symptoms. There were no recent travels away from home, and no animals at home.

The review of systems was significant for sore throat, generalized fatigue, fever, chills, and sweats, as noted above. He also had burning of the feet.

He had no known drug allergies. His medications before admission were diltiazem (extended-release form) 300 mg daily, started 3 months earlier, omeprazole 20 mg BID and sucralfate 1 g TID.

One day after admission, the infectious diseases consultant was asked to address this fever of unknown origin.

On examination, the patient was alert and oriented, and in no acute distress. The vital signs were as follows: BP 140/84, RR 24, HR 71, temperature range 102.5–103.6 °F; height 5′11″, weight 224 pounds.

Head and neck exam showed bilateral tonsilar exudate without erythema, otherwise normal. The heart, lung, abdomen, skin, genitalia, and neurologic examinations were unremarkable. The extremities were unremarkable except for dark toenails, consistent with onychomycosis.

Hospital course

The work-up that had been started before admission was continued. Consultations with rheumatology and hematology services were placed. Multiple blood cultures, chest and sinus x-rays, including CT scan of the chest, were negative (including for adenopathy). Urine and stool studies, including culture, ova, cysts and parasites, were negative. HIV, EBV, CMV, herpes, and Lyme disease serology were negative. Hemoglobin electrophoresis, antinuclear antibody tests, and angiotensin converting enzyme levels were either negative or in the normal range. The ESR was elevated at 54 mm/hour. CBC showed WBC 9.7, H/H 12.0/35.2, respectively, and platelet count 145,000. The differential count showed 67% polymorphs, 18% lymphocytes, and 14% monocytes. Chemistries, including liver tests, were normal. A purified protein derivative (PPD) skin test for tuberculosis was negative, with normal *Candida* control.

One day after the ID consult, and in the setting of a clinically stable patient, a decision was made to discontinue the antihypertensive drug started 3 months earlier (diltiazem). To manage or control the blood pressure in the interim, topical (patch) nitroglycerin was used instead. Twenty-four hours later, the fever resolved. Twenty-four hours after diltiazem was discontinued, ceftriaxone (the antibiotic which the patient had been on since admission) was also discontinued. He remained without fever for the next 2 days, and was subsequently discharged to outpatient follow-up. The temperature graph is shown in Fig. 2.3a.

Case discussion

This patient had a 2-month history of fever, chills, malaise, and sweats. Again, the extensive work-up, including cultures and scans, revealed no obvious cause for the fever. More detailed history and examination, however, suggested the possibility of drug-induced fever. He looked relatively well in between the fevers. His vital signs, especially the pulse rate (range: 69–94), remained relatively normal, even when the temperature was as high as 103.6 °F, while the respiratory rate remained in the normal range all through the recorded period. All the three medications he was on at the time of admission had been started in the 3 months prior to admission. Since drug-induced fever can occur even after months or years on a medication, it was a clinical "educated guess" decision to stop the diltiazem,

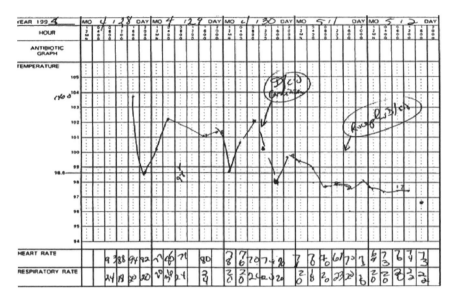

Figure 2.3a Graph of temperature and other vital sign recordings: 4/28–5/2/98.

based on the above noted considerations. Because he needed to be on medication to control his hypertension, we had to find an alternative drug, from a completely different class, to replace the one being withdrawn, hence the nitroglycerin patch. He had been switched to the diltiazem from a previous medication 3 months earlier for better control of his BP (he did not recall the name of the previous medication).

The rapid resolution of the fever after withdrawal of the "offending" drug rules out other possible disorders, including connective tissue disease, as a likely explanation. The response (fever resolution) was sustained. In both instances (the two cases of drug fever), the offending agents were subsequently listed in the patients' drug allergy list and records.

Brief review and discussion of drug-induced fever

Drug-induced fever (DIF) has been defined by many authors. Mackowiak and LeMaistre in 1987 defined drug fever as "fever coinciding with the administration of a drug and disappearing after discontinuation of the drug, when no other cause for the fever could be ascertained after a careful physical examination and appropriate laboratory study" [1].

The consequences of not arriving at a prompt diagnosis of DIF include exposing the patient to potentially hazardous treatment, as well as prolonged hospitalization. The extensive investigations needed to exclude other serious illnesses invariably lead to high cost for the patient and the institution.

Although the diagnosis of DIF is often one of exclusion, there may be subtle clues that can be obtained from a detailed and comprehensive history and physical examination. Johnson and Cunha in their 1996 review addressed some of the clinical features of drug fevers [2]. They point out that some individuals may be atopic or may have been on "sensitizing medication" for days or years.

The first of our patients was on risperidone for 4–5 days, while the second patient was on diltiazem for 3 months. We have no information for either of these patients about any prior sensitization. As also suggested by Johnson and Cunha, the fever may be low to high grade, patients often look "inappropriately well" for the degree of fever, and have relative bradycardia with temperature greater than 102 °F [2]. Rash is uncommon but when present is usually maculopapular and central.

The two patients presented here had high spiking fevers and did not look acutely ill or "septic." Both had relative bradycardia or inappropriately normal heart rates relative to the height of their spiking fevers, and neither had a rash. Results of laboratory tests may be variable, but normal values do not preclude a drug-induced fever.

Roush and Nelson [3], among many others [4,5], have described some of the mechanisms of DIF. These include factors that relate to **drug administration** (intrinsic pyrogenic activity, contaminated products, infusion or injection-related effects, e.g. amphotericin B and bleomycin), **idiosyncratic** (inherited genetic defect or unknown, e.g. volatile anesthetics, succinylcholine), **hypersensitivity** (immunologically mediated, e.g. methyldopa, phenytoin, procainamide), **pharmacologic** (release of endotoxin from killed organisms, tissue injury, e.g. penicillin and antineoplastic agents), and **alteration of thermoregulation** (peripheral vasoconstriction, increase in basal metabolic rate, and reduced perspiration, e.g. cocaine, amphetamine, levothyroxine, atropine, etc.).

The review by Mackowiak and LeMaistre addressed the temporal relationship between administration of drug and onset of fever [1]. They pointed out that antineoplastic drugs have a shorter median lag time (0.5 days) than any other agent. The median number of days for cardiac drugs was 10, for antimicrobials 6, and for central nervous system drugs 16 days. However, the means vary and the standard deviation may be wide. Previous exposure or sensitization would clearly shorten the noted lag time to fever (Table 2.3a).

A comprehensive review of temperature regulation and the pathogenesis of fever has been presented by Mackowiak elsewhere [6].

Lessons learned from these two cases

- Drug-induced fever (DIF) is typically a diagnosis of exclusion.
- Any drug may be responsible for DIF.
- Drug-induced fever can occur early or late in the course of drug therapy.

Table 2.3a Temporal relationship between administration of drug and onset of fever.

Class of Offending Agent	Episodes No.	Lag Time (Days)		
		Mean	Median	SD
Cardiac	36	44.7	10	131.1
Antimicrobial	44	7.8	6	8.4
Antineoplastic*	11	6.0	0.5	12.3
Central Nervous System	24	18.5	16	15.4
Other	20	18.8	7	34.1

*Significantly shorter median lag time than any other agent.

SD, standard deviation.

Adapted from Mackowiak and LeMaistre [1]. Reproduced with permission from the American College of Physicians. © 1987 The American College of Physicians.

- A thorough drug history and patient observation can be useful in reaching a diagnosis.
- The patient may look inappropriately well in spite of high fever in DIF.
- Relative bradycardia or inappropriately normal heart rate may be a useful clue in DIF.
- Work-up of DIF can be very expensive, and hospitalization prolonged.
- The most important clue in the diagnosis of DIF is thinking about it in the differential diagnosis of fever of unknown origin.

References

1 Mackowiak PA, LeMaistre CF. Drug fever: a critical appraisal of conventional concepts. Annals of Internal Medicine 1987; 106: 728–33.
2 Johnson DH, Cunha BA. Drug fever. Infectious Disease Clinics of North America 1996; 10(1): 85–91.
3 Roush MK, Nelson KM. Understanding drug induced febrile reactions. American Pharmacy 1993; NS33(10): 39–42.
4 Tabor PA. Drug-induced fever. Drug Intelligence and Clinical Pharmacy 1986; 20: 413–20.
5 Lipsky BA, Hirschman JV. Drug fever. JAMA 1981; 245(8): 851–4.
6 Mackowiak PA. Temperature regulation and the pathogenesis of fever. In: Mandell GL, Bennett JE, Dolin R (eds) *Principles and Practice of Infectious Diseases*, 7th edn. Philadelphia: Churchill Livingstone, 2010, pp. 765–78.

Further reading

Mackowiak PA, Durack D. Fever of unknown origin. In: Mandell GL, Bennett JE, Dolin R (eds) *Principles and Practice of Infectious Diseases*, 7th edn. Philadelphia: Churchill Livingstone, 2010, pp. 779–89.

CHAPTER 3

Dermatologic Manifestations of Infectious and Non-infectious Diseases

Case 3.1 A 45-year-old patient referred for "persistent shingles"

A 45-year-old Caucasian female was admitted for severe neck pain in March, 2010 for possible surgery to relieve symptomatic disk herniation of the C5–6 area. However, evaluation for surgery was postponed because of a chronic persisting rash in the right inner gluteal area over the previous 3–4 months. This rash was thought to be due to shingles, but had been unresponsive to antiviral (valaciclovir, aciclovir) and topical steroid (triamcinolone) therapy.

Infectious diseases consult was sought 1 day after admission, for advice and clearance before a planned cervical surgery for relief of the neck pain.

Three and a half months before admission, the patient had noted a "ringworm-like" rash about the size of a quarter that grew over the next month to several centimeters in size, located in the right inner buttocks and perineal area. It was itchy and burning, and caused the patient to scratch, worsening the burning of the irritated skin. Antiviral therapy given systemically over 3 months (aciclovir, valaciclovir) had not helped, nor had topical steroid therapy.

She had no similar rash elsewhere. She had no fever or lymphadenopathy in the groin area. She continued to have severe pain in her neck and slight weakness of her legs, left more than right.

The past medical history was significant for type 2 diabetes mellitus, multiple previous kidney stones, even requiring a temporary stent 2 years earlier in the right ureter, and a left breast CA, treated with lumpectomy and subsequent chemotherapy and radiation therapy 3 years earlier. Other medical and surgical conditions included degenerative joint disease: C6–7 fusion in 2007; lumbar laminectomy at L5 8 years earlier; and hysterectomy in 2006.

The patient was married, had two adult sons; and a strong family history for cardiovascular disease in male relatives. She did not smoke or drink.

On examination, she was alert and oriented, and in no acute distress. Vital signs were as follows: blood pressure (BP) 114/66, respiratory rate (RR) 16, heart

Cases in Clinical Infectious Disease Practice: Obtaining a Good History from the Patient Remains the Cornerstone of an Accurate Clinical Diagnosis: Lessons Learned in Many Years of Clinical Practice, First Edition. Okechukwu Ekenna.
© 2016 John Wiley & Sons, Inc. Published 2016 by John Wiley & Sons, Inc.

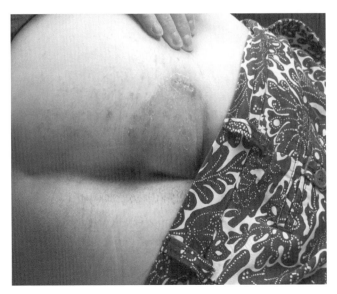

Figure 3.1a Eczematous skin changes in the right buttock-perineal area. Photo taken on 4/8/10, 9 days after treatment was started. The indurated area above is the site of a skin biopsy on 3/31/10 (reproduced with permission).

rate (HR) 97, temperature 98.4 °F; height 5′4″, weight 158 pounds. Head and neck were significant for a prominent torus palatinus. Heart, lung, abdominal, and extremity exams were unremarkable. There was no palpable adenopathy. The neurologic exam showed only a slight weakness of the left leg, compared to the right.

The skin was remarkable only for the eczematous changes noted in the right inner buttock, ischial, and perineal areas with "ringworm-like" changes, with peripheral excoriations and some "central clearing." This 5–6 cm relatively dry lesion had no pustules, no active blisters noted, and no surrounding cellulitis. A photograph of the skin lesion taken 9 days after treatment was started is shown in Fig. 3.1a.

Initial laboratory tests

Routine initial laboratory findings were unremarkable.
- *What is your initial impression and what tests would you want to order at this time?*

A differential diagnosis was generated. Skin scrapings were obtained for dermatophyte (superficial fungal) culture, and skin punch or excisional biopsy was requested. The patient was then started on presumptive or tentative therapy the same day. After the skin biopsy the next day, the patient was discharged to outpatient follow-up.
- *What are the likely differential diagnoses in this patient?*
- *What treatment would you start this patient on?*

Follow-up outpatient visits

The patient was seen 8 days after discharge (9 days after the initial hospital consult) in the office. The eczema was less intense; the biopsy site showed no sign of infection or purulence. A photograph was taken, as shown in Fig. 3.1a. Culture report of the skin scraping was negative. She was continued on the same therapy.

On return office visit 21 days later (4/21/10), most of the eczema had cleared. The final culture report was still not available. The patient clearly was feeling better with respect to the rash, but still had continuing neck pain.

A final visit for the rash occurred 5 weeks later on 6/3/10, at which time the eczema was completely resolved. She had stopped the prescribed medication 2 weeks earlier, with no recurrence. She was ready to be evaluated for the planned cervical neck surgery.

- *Have you now changed or modified your presumptive diagnosis?*
- *What is your final diagnosis before the results of the investigation are reviewed?*

Laboratory and histopathology results

The fungal preparations (10% KOH) of the skin biopsy and scrapings were negative. The histopathology was read as "dermatophytosis with excoriation and lichen simplex chronicus."

The dermatophyte culture of the skin scrapings obtained on 3/30/10 was first reported to be positive on 4/27/10, at 4 weeks. The final report was sent out on 5/5/10, 5 weeks after the initial culture, based on the lactophenol cotton blue stain shown in Fig. 3.1b.

Case discussion

This patient had an eczematous lesion in the right buttock and perineal area that was first treated for shingles (herpes zoster). It failed to respond (persisting after 3 months) to the usual treatment for herpes zoster or human herpes simplex virus (HSV). The characteristics of the eczema (rash) did not fit shingles or HSV. The lesion did not fit a typical dermatomal distribution for herpes zoster; it was eczematous, without grouped blisters or vesicles; it had some central clearing, was scaly, and very itchy. Those characteristics are more in line with a dermatophyte infection.

It was also not a simple eczema or contact dermatitis, because the severe itching did not respond to topical steroid therapy. Instead, it continued to get larger on antiviral or steroid treatments.

The nature of the lesion suggested a dermatophyte infection so we obtained skin scrapings for dermatophyte culture. Final culture results were not expected for 3–4 weeks, and so presumptive therapy was started immediately. Because of the prolonged nature of the lesion and the unusual location, a biopsy was requested to rule out the remote possibility of a malignancy. The histopathology did not show a malignancy or melanoma but rather a non-specific report (dermatophytosis with excoriation and lichen simplex chronicus) that did not influence our treatment recommendation.

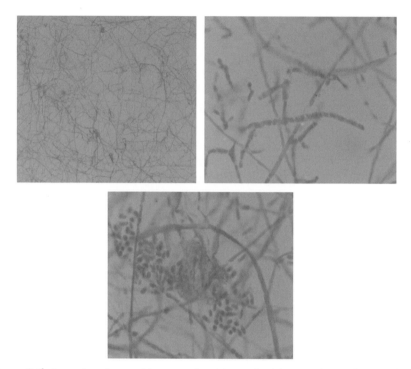

Figure 3.1b Lactophenol cotton blue prep of aerial growth of dermatophyte culture of skin scraping at 5 weeks. Septate hyphae and microconidia are seen on the slides.

The patient had been started on a topical over-the-counter antifungal regimen (Tolnaftate), and was already showing some clinical response 9 days after initiation of therapy. The photograph of the skin lesion in Fig. 3.1a was taken during the first follow-up visit in the office, 8 days after discharge from the hospital. By the last office visit 2 months later, all the eczema had cleared.

The source of the dermatophyte infection remains unclear, although the patient recalled the installation of a new bathtub in their home weeks or months before onset of the lesion. The natural habitat of dermatophytes may include humans, animals, and soil [1].

• *Final diagnosis: Dermatophyte infection due to Trichophyton species.*

Dermatophyte infections

Dermatophyte infections are common worldwide, and can lead to a variety of clinical manifestations that depend on the part of the body involved. Infections are usually named according to the anatomic location involved, e.g. tinea capitis, tinea corporis, and tinea cruris.

Dermatophytes are superficial, keratinophilic fungi that are capable of invading the keratinous tissues of living animals.

Three types of dermatophytes account for the majority of infections: *Epidermophyton*, *Trichophyton*, and *Microsporum*. The fungi attack skin, nails, and hair, where keratin is the major structural protein (epidermis), leading to a wide variety of disease states. Typically, healthy people are affected, although unusual presentations may be seen in immunocompromised patients or those with underlying immune defects or problems.

Diagnosis is usually made via KOH preparations of skin, nail or tissue; culture often takes several weeks for final identification in special dermatophyte medium or typically Sabouraud's dextrose agar (SDA) medium, incubated at 25–30 °F. Identification of the dermatophyte species is often based on the colony characteristics in pure culture on SDA, and microscopic morphology [1].

Lessons learned from this case
- Initial misdiagnosis of the rash led to a delay in definitive care of the neck pain.
- A good history and physical exam would have suggested the eczematous nature of the rash.
- Time and money were lost in the pursuit of expensive and ineffective treatment.
- A simple, cheap, and affordable over-the-counter medication was able to control the rash.
- Additional expensive work-up and hospital admission would have been avoided with earlier diagnosis of the nature of the rash.
- Anxiety was relieved with the knowledge that this was not a "herpes" infection, providing great emotional relief for the patient.
- Knowledge about the slow growth rate of dermatophytes in special culture media allowed for patience in carrying through the slow but effective topical treatment of the infection.

Reference

1 Summerbell RC. *Trichophyton, Microsporum, Epidermophyton*, and agents of superficial mycoses. In: Murray PR, Baron EJ, Jorgensen JH, Pfaller MA, Yolken RH (eds) *Manual of Clinical Microbiology*, 8th edn. Washington, DC: ASM Press, 2003, pp. 1798–819.

Further reading

Goldstein AO, Goldstein BG. Dermatophyte (tinea) infections. Available at: www.uptodate.com/contents/dermatophyte-tinea-infections?view=print, accessed February 25, 2016.

Centers for Disease Control and Prevention (CDC). Ringworm Information for healthcare professionals. Available at: www.cdc.gov/fungal/diseases/ringworm/health-professionals.html, accessed March 1, 2016.

Case 3.2   A 39-year-old female with fever and acute rash illness

A 39-year-old Caucasian female was admitted in early February, 2002 for persistent fever, chills, and generalized fatigue for 2 weeks. She also had flu-like symptoms of headaches and malaise, plus generalized weakness. Rash was noted for the 5 days preceding admission, and sore throat, plus mouth sores, including on the lips and buccal cavity, for 3–4 days.

Infectious diseases consult was sought for fever, rash, and flu-like symptoms on the day of admission.

The review of systems was positive for the following: rash, present on the head and face, especially trunk, and upper extremities more than lower extremities. The rash involved the palms and soles of the feet, and did not itch. The patient had generalized body and joint aches, especially knees.. Bumps were noted in her buttocks area and she had occasional productive cough, but was a smoker.

The past medical history was significant for the following: in June 2001 she was admitted for 3 days for acute appendicitis with epiploica. Surgery was done by laparoscopy; histopathology showed chronic periappendicitis with focal serositis. Tubal ligation was performed several years earlier, and she had tonsillectomy as a child.

She was on no medications prior to this admission and was allergic to sulfa (nausea and vomiting).

Family and social history was significant for the following: she was single but had three children in their early 20s. She admitted to being a prostitute, with several partners per day and in the profession for more than a year. Her clients sometimes did not use condoms. She previously lived and did business in a neighboring state. She had significant alcohol intake: up to 3–4 six-packs of beer per day, as well as using cocaine occasionally. She had smoked about one pack of cigarettes/day for the last 20 years.

The physical exam showed her to be alert, oriented, but with depressed facies; she was chronically ill-looking but not in acute distress.

Her vital signs were as follows: BP 103/55, RR 20, HR 96, temperature 101.1 °F; height 5'1", weight 102 pounds.

The head and neck exam showed scleral jaundice, dental caries, and super-ficial erosions in the left upper lip as well as faint lesions on the tongue. There was no thrush. The face also showed macular papular rash (as elsewhere). There was no adenopathy in the neck. The heart and lung exam was unremarkable. Abdomen was soft, with fullness of the right upper quadrant and tenderness to palpation. The liver was palpable and tender, while the spleen was not palpably enlarged. The examination of the external genitalia showed no vaginal discharge, but no speculum exam was done. There was a ruptured blister noted on both sides of the medial gluteus below the vulva (perineal/perirectal area).

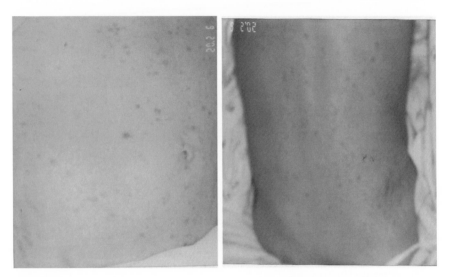

Figure 3.2a Maculopapular rash on the abdomen and back: photo taken on 2/9/02.

Lymph nodes were prominent only in the right inguinal, but not in the axilla or neck. Except for the rash, the examination of the upper and lower extremities was normal. Neurologic exam was normal, except for the depressed facies.

The skin showed generalized macular papular rash, especially on the trunk anteriorly and posteriorly, more than elsewhere; in the upper more than lower extremities; and involving the soles and palms, as well as the face (see Fig. 3.2a).

Laboratory and radiologic findings

The chest x-ray showed no acute pulmonary disease. The complete blood count (CBC) showed a white blood cell (WBC) count of 11.6, hemoglobin and hematocrit (H/H) of 13.5/38.9, respectively, mean corpuscular volume (MCV) 91.5, and platelet count 270,000. The differential count showed 69% polymorphonuclear leukocytes (PMNs), 2% bands, 20% lymphocytes, and 7% monocytes. Erythrocyte sedimentation rate (ESR) was 21 mm/hour, prothrombin time was elevated at 40.6 seconds. Urine drug screen was positive for cocaine while the urinalysis was otherwise normal. The chemistries showed a sodium value of 135, potassium 4.0, glucose 106 mg/dL, BUN and creatinine 6 and 0.5, respectively. Liver tests showed levels of serum glutamic oxaloacetic transaminase (SGOT) of 255, serum glutamic pyruvic transaminase (SGPT) 896, alkaline phosphatase 283, total bilirubin 4.98 (conjugated 3.06, unconjugated 0.56, delta 1.30), albumin 2.7, and total protein 6.2 g/dL. See Table 3.2a for serial liver test results.

Blood cultures ×2 were negative. Acute hepatitis serology: reactive (+) for HBsAg and HB core Ab IgM. The rectal blister lesion was cultured for viruses.
- *What is your presumptive diagnosis?*
- *What are the differential diagnoses to consider?*
- *What additional laboratory tests would you want to order?*

Table 3.2a Results of serial liver panel tests.

Test	2/11/2002	2/09/2002	Normal Value
SGOT	186	255	15–46 IU/L
SGPT	457	896	11–66 IU/L
Alk Phos	270	283	38–126 IU/L
Conjugated Bilirubin	0.63	3.06	0.00–0.03 mg/dl
Unconjugated Bilirubin	0.41	0.57	0.00–1.10 mg/dl
Delta Bilirubin	1.56	1.30	0.00–0.45 mg/dl
Total Bilirubin	2.61	4.98	0.2–1.3 mg/dl
Albumin	2.0	2.7	3.1–5.1 g/dl
Total Protein	5.3	6.2	6.3–8.2 g/dl

SGOT, serum glutamic oxaloacetic transaminase; SGPT, serum glutamic pyruvic transaminase.

Differential diagnoses

The major differential diagnoses considered before the final laboratory test results were available included the following: secondary syphilis, acute HIV infection, acute hepatitis (alcohol, substance use, viral, bacterial or other acute illness), or a combination of any of the above.

Hospital course, additional laboratory results, and clinical findings

The rapid plasma reagin (RPR) test for syphilis was reported positive at 1:32 titer the next day. HIV serology and viral (HSV) culture of the rectal blister fluid results were still pending. Additional physical examination showed lesions on the clitoris and vulva, consistent with the other skin rash noted elsewhere on the body.

A decision was made to begin treatment with benzathine penicillin (Bicillin L-A), but to be given the next day, so that the patient could be better observed for possible reaction to treatment by the morning nursing shift.

The local health department was contacted for follow-up and contact tracing.

Treatment and observation findings in the hospital

The next morning, on 2/11/02, at 0600 hours the patient received 2.4 M units of benzathine penicillin by intramuscular injection. Her temperature rose from 101.1 °F to 101.8 °F at 0800 hours, and then peaked at 102.8 °F at 1600 hours the same day (see Fig. 3.2b).

By the next morning on 2/12/02, the fever had resolved. The patient was seen by the local health department personnel, and outpatient follow-up and treatment completion (a total of weekly doses of 2.4 M units of benzathine penicillin ×3) were planned, as well as contact tracing arranged, before formal discharge later that day.

• *What phenomenon or reaction is shown in Fig. 3.2b?*

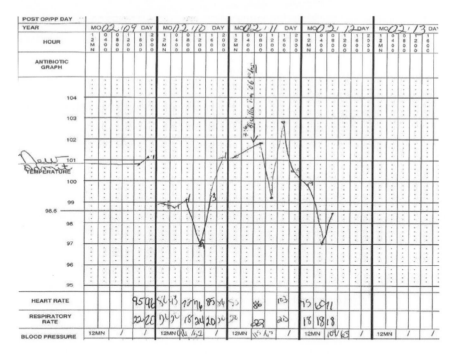

Figure 3.2b Graphic display of temperatures before and after penicillin treatment.

Case discussion

A more detailed case discussion will follow after the next case in this section. However, the phenomenon shown on the temperature graph is presumptive Jarisch–Herxheimer reaction [1].

The cause of this reaction is not completely understood, but the release of *Treponema pallidum* lipoproteins with inflammatory activities from dead or dying organisms is the likely inducer of this clinical phenomenon. This is thought to be a reaction to endotoxin-like products released by the death of micro-organisms (in this case, *T. pallidum*) within the body during antibiotic treatment. Also described as "therapeutic shock," the reaction is a local and systemic exacerbation of whatever stage of syphilis is being treated: reactions occur within 12 hours of initiation of treatment, and are usually over by 24 hours.

This phenomenon was first observed by Jarisch (1895) and Herxheimer (1902) with the use of mercury for treatment of syphilis, but it has now been described with other treatments as well. This febrile reaction is most prominent during treatment of secondary syphilis because of the high quantity of circulating antigens.

Antipyretic and analgesic treatments are useful in reducing the symptomatic discomfort of this phenomenon (fever, chills, body aches, etc.). This patient received acetaminophen for relief of symptoms. She was feeling much better at the time of discharge home 1 day after initiation of penicillin treatment.

In addition to secondary syphilis, the patient had acute hepatitis, with elevated transaminases. The total and conjugated bilirubin came down relatively rapidly. The hepatitis likely reflected a combination of acute hepatitis B infection (positive HBsAg and IgM), as well as effects of alcohol and other drug use, plus the secondary syphilis. It would be difficult to tease out the influence of each of these components on the liver test results. HIV serology was negative, as was the rectal blister lesion viral culture.

- *Final diagnosis: Secondary syphilis, with associated liver disease.*

Reference

1 US Department of Health, Education, and Welfare. *Syphilis: A Synopsis*. Public Health Service Publication No. 1660. Washington, DC: US Government Printing Office, 1967.

Case 3.3   A 45-year-old female with subacute rash illness for 2 weeks

In early January, 2011, a 45-year-old Caucasian female was admitted with a 2-week history of rash. The rash had started on the chest, then spread to involve the rest of the body including the scalp, face, hands, and feet. The rash was itchy, and so the patient thought it was an allergic reaction.

Six days before admission, she came to the emergency room (ER) with this papular urticarial rash, along with mouth ulcerations, and was prescribed Medrol dose pack (methylprednisolone), tramadol, and trimethoprim/sulfamethoxazole. She had no relief from these, and so returned on 1/12/11 to the ER, and was admitted.

The review of systems was positive for pruritic rash ×2 weeks. The rash extended from the edge of the scalp to the face, torso, and extremities, palms and soles included. There were mouth ulcerations which were associated with pain on swallowing. She did not have a high fever, but had a low-grade temperature and subjectively felt feverish.

Infectious diseases consult for generalized rash was placed late on 1/13/11 but a full (physical) evaluation was done 2 days later.

The past medical history was significant for hypertension for many years; obesity (peak weight 400 pounds, now 283 pounds); chronic stable asthma since childhood; seizure disorder for years (last seizure was 3 months earlier); and a history of depression, with emotional lability. She had previous treatment for hepatitis C viral (HCV) infection >10 years ago (allegedly with PEG interferon); cholecystectomy and appendectomy 15 years ago; and hysterectomy more than 10 years earlier.

The family and social history was significant for being separated from her boyfriend, although they still saw each other occasionally. She had three children in their 20s, living away from home. Family history was positive for heart and lung disease, as well as diabetes in several family members. Although there was previous alcohol use, she claimed none in years, as well as previous cocaine use in the past. She was unable to say the last time she had any illicit drug use.

Allergies were to codeine, acetaminophen, propoxyphene (nausea and vomiting), morphine or dihydrocodeine, acetaminophen and caffeine combination (throat closure/anaphylaxis), and penicillin (hives/anaphylaxis).

Medications prior to this admission included phenobarbital, phenytoin, methylprednisolone, trimethoprim/sulfamethoxazole, occasional acetaminophen and hydrocodone, and alprazolam, on as-needed basis.

The last admitted sexual intercourse was 6 months earlier, although she was "separated" from her boyfriend. She denied any sexual promiscuity otherwise. The boyfriend was said to be a substance user (alcohol). The rash that started in the chest area had spread within 1 week to involve the other parts of the body.

On examination, she was alert, oriented, tired, depressed, and emotionally labile. She was obese and in no acute distress.

The vital signs were as follows: BP 146/84, RR 20, HR 87, temperature 95.9 °F; height 5′8″, weight 283 pounds.

Head and neck exam showed no conjunctivitis, but some puffiness around the eyes and face with rash. There was coating and ulcerations of the buccal cavity, especially of the tongue. The heart, lung, and abdomen showed nothing significant except the large abdominal pannus and rash. Lymph nodes were not palpably enlarged in the neck, axilla, or groin. The extremities showed a predominance of the rash in the extensor more than flexor areas, the upper more intensely than in the lower extremities. The skin showed a generalized maculopapular rash, partially blanching, intense in the back and anterior upper chest, less intense on the abdomen. Various forms, from squamous and peeling to maculopapular and even pustular, were noted (Figs 3.3a, 3.3b).

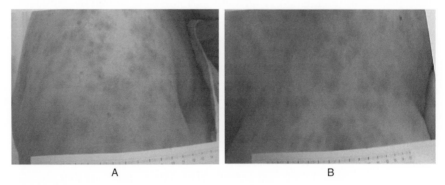

A   B

Figure 3.3a A. Right upper arm and shoulder: maculonodular lesions, some scaling noted. B. Anterior chest lesions: maculonodular, only partially blanching. Photos taken on 1/13/11 (reproduced with permission). (*See insert for color information*)

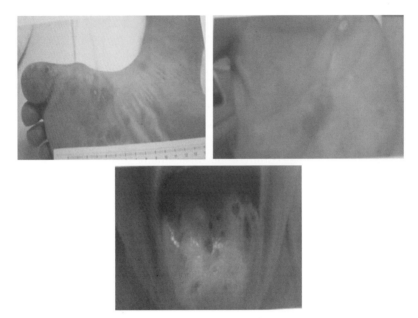

Figure 3.3b Right foot: plantar desquamative erythematous lesions; pustular right palmar lesions, with surrounding erythema; and ulcerative lesions on the tongue. Photos taken 1/13/11 (reproduced with permission).

- *What is your diagnosis or differential diagnoses?*
- *What other laboratory tests would you need to make or confirm the diagnosis?*
- *Justify your choice of tests and diagnoses.*

Laboratory findings

Blood cultures ×2 were negative. CBC showed a WBC count of 6.2, H/H 13.6/40.6, MCV 98.5, and platelet count of 168,000. The differential counts were as follows: 66% PMNs, 17.7% lymphocytes, 13.8% monocytes, and 2.4% eosinophils. The sedimentation rate was 30 mm/hour. Urinalysis and culture were negative; urine drug screen was positive for cocaine. Chemistries showed a sodium level of 137, potassium 4.5, glucose 89, BUN and creatinine 14 and 1.0, respectively. Liver function tests were normal, except for a low albumin of 2.7 mg/dL. Rheumatoid factor was negative. The phenytoin and phenobarbital levels were low. Urine paraproteins and *Mycoplasma* screens were negative.

- *Has your differential diagnosis changed, based on the above laboratory data?*

Additional laboratory tests

Syphilis IgG Antibody screen was positive (>1.11 index); the requested titer was not done. This antibody test had recently replaced the previous RPR test used in our hospital as a screening test. The *T. pallidum* total antibodies value was 20 (reference range <5 units). This test is the equivalent or replacement for the FTA-ABS quant (units are reported as fluorescent intensity units).

N.B. In another patient seen in October, 2010, the syphilis IgG Ab screen of 2.91 was the equivalent of a RPR titer of 1:256 in serum samples drawn 1 week apart.

HCV titer was 1,758,189 IU/mL (log 6.25) but cryoglobulin was not detected at 72 hours or on day 7. HIV serology was negative.

Presumptive diagnosis

Our working diagnosis was secondary syphilis.

The patient's HCV infection was diagnosed more than 10 years ago. It was determined on this admission that the HCV was subtype 1a, with a high viral load. It is therefore doubtful that she had adequate therapy then, or she may not have had a sustained viral response. It is also unclear how well she may have adhered to the prescribed therapy in the past.

This patient was allergic to penicillin (anaphylaxis), and therefore doxycycline was prescribed instead. The health department was contacted for outpatient follow-up of the patient, along with the primary care physician, as well as for the relevant contact tracing and other monitoring.

She was seen 2 years later in 2013 for an unrelated problem, and admitted to completing her treatment with doxycycline through the local health department. She had no rash at that time.

• *Final diagnosis: Secondary syphilis.*

Case discussion of the two patients with secondary syphilis (cases 3.2 and 3.3)

The patient in case 3.3 had secondary syphilis, like the preceding case. However, this patient also had HCV infection, with high viral load. Fever was not prominent in this patient, unlike in the previous patient, who had a 2-week history of fever but a shorter acute onset of rash (4–5 days), before admission. The first patient had acute HBV infection and abnormal liver tests while the second patient had chronic HCV infection and a more prolonged period of rash. Interestingly, also for the second patient, the rash was pruritic (but not for the previous patient), something that is not very typical for secondary syphilis. The first patient had the more classic presentation, with all the systemic symptoms of malaise, body aches, fever, arthralgias, etc. She also had the Jarisch–Herxheimer reaction to penicillin therapy, with high fever [1].

The rash was, however, more prominent and persistent in the patient with chronic HCV infection. It is possible that some of the symptoms reflected extrahepatic or autoimmune manifestations from chronic HCV infection [2]. However, cryoglobulin was not detected in this patient and there was no clinical evidence for porphyria cutanea tarda. There was also no significant adenopathy noted in either of the two patients reported here, except in the inguinal area for the first patient.

General comments on syphilis

The key to diagnosis of syphilis remains a high index of suspicion. Non-treponemal (NT) tests (VDRL, RPR card test) are useful for screening, while treponemal tests (FTA-ABS, MHA-TP) are used to confirm the diagnosis. False-positive reactions can occur in many disease conditions with the NT tests but are rarer with treponemal tests.

Definitions

- Early syphilis is defined as the stages of syphilis (primary, secondary, and early latent) that typically occur within the first year after acquisition of the infection. Early syphilis implies infection within 1 year.
- Latent syphilis is characterized by asymptomatic infection with a normal physical examination, in association with a positive serology.
- Late syphilis is arbitrarily defined as the stages of syphilis that occur after early or early latent syphilis. Latent syphilis of uncertain duration is managed as if it were late latent syphilis.

Clinical manifestations

In primary syphilis, the classic lesion is a painless ulcer associated with mild-to-moderate regional lymphadenopathy. Secondary syphilis patients have marked systemic symptoms, such as malaise, fever, headache, and rash. The stage of the disease has implications for treatment, choice of agent, and duration of therapy.

Reported causes of false-positive tests for syphilis [3]

Non-treponemal tests (VDRL, RPR) – acute

- Pneumococcal pneumonia; scarlet fever; infective endocarditis
- Leprosy; lymphogranuloma venereum; relapsing fever; malaria
- Rickettsial infections; psittacosis; leptospirosis; chancroid
- Tuberculosis; *Mycoplasma* infections; trypanosomiasis; *Varicella* infections
- HIV; measles; infectious mononucleosis; mumps; viral hepatitis; pregnancy

Non-treponemal tests (VDRL, RPR) – chronic

Chronic liver disease; malignancy (advanced); injection drug use; myeloma; advanced age; connective tissue disease; and multiple transfusions.

Treponemal tests (FTA-ABS, MHA-TP)

Lyme borreliosis; leprosy; malaria; infectious mononucleosis; relapsing fever; leptospirosis; and systemic lupus erythematosus.

Lessons learned from these cases

- Syphilis is a protean disease, and can manifest in many different ways, especially in persons with other medical conditions.

- Secondary syphilis typically presents with systemic symptoms, often including malaise, fever, and body aches.
- The typical rash of secondary syphilis involves the palms and soles, and may involve the mucous membranes.
- Serologic tests are necessary to separate syphilis from other systemic illnesses.
- The Jarisch–Herxheimer reaction is an expected clinical response to effective therapy, reflecting release of *T. pallidum* lipoproteins, with inflammatory activities from dead or dying organisms, causing fever and body aches.

References

1 US Department of Health, Education, and Welfare. *Syphilis: A Synopsis*. Public Health Service Publication No. 1660. Washington, DC: US Government Printing Office, 1967.
2 Chopra S, Flamm S. Extrahepatic manifestations of hepatitis C viral infection. Available at: www.uptodate.com/contents/extrahepatic-manifestations-of-hepatitis-c-virus-infection?view=print, accessed February 25, 2016.
3 Hicks CB. Diagnostic testing for syphilis. Available at: www.uptodate.com/contents/diagnostic-testing-for-syphilis?view=print, accessed February 25, 2016.

Further reading

Chopra S, Muir A. Treatment regimens for chronic hepatitis C virus genotype 1. Available at: www.uptodate.com/contents/antiviral-therapy-for-chronic-hepatitis-c-virus-genotype-1?view=print, accessed February 25, 2016.
Hicks CB, Sparling PF. Pathogenesis, clinical manifestations, and treatment of early syphilis. Available at: www.uptodate.com/contents/pathogenesis-clinical-manifestations-and-treatment-of-early-syphilis?view=print, accessed February 25, 2016.
Singh AE, Romanowski B. Syphilis: review with emphasis on clinical, epidemiologic, and some biologic features. Clinical Microbiology Reviews 1999; 12(2): 187–209.
Sparling PF, Hicks CB. Pathogenesis, clinical manifestations, and treatment of late syphilis. Available at: www.uptodate.com/contents/pathogenesis-clinical-manifestations-and-treatment-of-late-syphilis?view=print, accessed February 25, 2016.

Case 3.4   An 82-year-old man with a chronic non-healing elbow wound

An 82-year-old Caucasian male first noted a bulge or growth in the medial left antecubital area of his arm in September 2008. It was just a swelling, and he thought nothing of it initially. He was not one to worry about these things, and "toughed it out" for a while.

His family convinced him to have this evaluated. He had an incision and drainage procedure done on 1/2/09, with pus drained from the site. He subsequently received outpatient wound care elsewhere, where he was seen twice a week. Over the next 2.5 months he received various antibiotics including clindamycin (which caused him GI symptoms), trimethoprim/sulfamethoxazole, amoxicillin/clavulanate, and most recently a combination of trimethoprim/sulfamethoxazole and rifampin. The cultures obtained from the wound in February and March, 2009 showed methicillin-sensitive *Staphylococcus aureus* (MSSA). A bone scan done on 3/6/09 was suspicious for osteomyelitis.

He was referred to my office on 3/18/09 because of the non-healing wound, post surgical debridement 2.5 months earlier, because it was unresponsive to antimicrobial therapy.

The review of systems was significant for lack of any fever. The wound was still draining, and would not heal for several months. There was swelling at the site of the wound and in the left arm. The patient had no known drug allergy.

The past medical history was significant for the following: hypertension for 12 years; arthritis of the right shoulder; left inguinal hernia repair; chronic swelling of lower extremities, for which he was supposed to wear compression stockings. He had a previous abdominal surgery (suprapubic scar), but was unsure what for. A diagnosis of non-Hodgkin's lymphoma was made 4–5 years ago. He had been seen by an oncologist, but had not received any specific therapy for the lymphoma.

He was widowed, had four adult children, and retired after holding many odd jobs, including working as a carpenter. He was a former heavy smoker (2–3 packs of cigarettes/day for 50 years), but stopped smoking 15 years earlier. A former heavy drinker for years, he had not drunk alcohol in 6 years. Family history was significant for hypertension and undefined cancer.

On examination, he was alert and oriented, pleasant, and in no acute distress. His vital signs were as follows: BP 150/84, RR 20, HR 100, temperature 100.0 °F; height 5'10", weight 181 pounds. Head and neck exam showed prosthetic dentures, but nothing else significant. Chest showed increased anteroposterior (AP) diameter, with lungs showing diminished breath sounds. The heart exam was unremarkable and the abdomen was normal, except for healed scars in the left inguinal and suprapubic areas from previous surgeries. Lymph nodes were not palpably enlarged in the neck, right axilla, or groin but in the left axilla there were two palpable clinically enlarged nodes. The lower extremities showed 3+ pitting edema, and changes consistent with postphlebitis syndrome. The upper extremities showed large hands and evidence for osteoarthritis. The left arm had swelling of the forearm and also of the distal upper arm, especially medially. Photographs of the left arm are shown in Fig. 3.4a.

• *What is your working diagnosis at this stage, and what would you do next?*

Office impressions and recommendations

The patient had a left medial elbow chronic abscess and cellulitis, with ulceration, that had been present for months, without resolution. In spite of appropriate

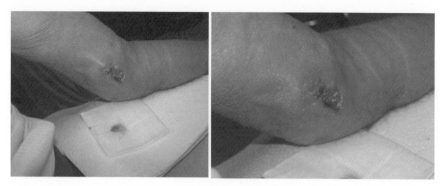

Figure 3.4a Left arm and elbow edema with inflammatory abscess: note the bluish area of the wound with abscess at the edge. Office photo taken on 3/18/09 (reproduced with permission). (*See insert for color information*)

antimicrobial therapy for the isolated very sensitive MSSA, the patient was not better, hence the referral. I noted in addition two palpable left axillary lymph nodes on exam. Although one bone scan had suggested osteomyelitis, this was an unsatisfactory explanation for the persistence of the edema for several months. Clearly, the lymphatic drainage was impaired clinically. Besides, there was this past history of untreated non-Hodgkin's lymphoma.

The patient's relative (who accompanied the patient) showed me photographs of the arm taken in December, 2008 and January, 2009. The photos showed a bulging mass that appeared cystic before the initial surgical incision.

The recommendation was made to admit the patient, and to surgically obtain tissue for histopathology, as well as for routine gram stain and cultures, including for fungal and acid-fast stains and cultures. Meanwhile, pus was easily obtained by gentle pressure from the edge of the bluish inflammatory area shown in the photograph, for microbiologic studies.

Follow-up events

The patient was not admitted to the hospital until 1 week later. The infectious diseases consultant was reconsulted. The patient had lost 7 pounds since the office visit 1 week ago, but the rest of the exam was not much changed, except the left arm. There was evidence for some tunneling in the area of the abscess in the medial elbow and in addition, multiple "satellite" lesions suggesting progression of the abscess. He still had no fever (temperature was 97.0 °F) while the rest of the vital signs were stable.

Laboratory and radiologic findings included the following: on 3/25/09, the CBC showed a WBC count of 3.5, H/H 9.8/29.3, respectively, MCV 98.7, and platelet count 148,000. The differential count showed 72% PMNs, 27.4% lymphocytes, and 0.6% monocytes. Chemistries showed a sodium value of 136, potassium 5.5, glucose 126 mg/dL, and BUN/creatinine 13 and 0.9, respectively. Cultures obtained from the left elbow abscess prior to this admission on 1/2/09,

1/26/09, 2/12/09, and from my office on 3/18/09 were all positive for MSSA. C-reative protein on 1/28/09 was 1.738, and sedimentation rate was only 8 mm/hour. Magnetic resonance imaging (MRI) of the left arm done on 1/23/09 described an inflamed enhancing mass, like a multiple, irregularly shaped abscess in the left elbow, with extensive edema. None of these tests, including the bone scan of 3/6/09, were specific about the etiology of the chronic wound and swelling.

The patient was taken to surgery on 3/28/09 for biopsy and debridement of the left arm. At surgery, spongiform necrosis was found within the skin and subcutaneous tissue, extending into a large necrotic defect near the epicondyle of the humerus. A large, locally invasive, poorly differentiated squamous cell carcinoma, arising in the medial left elbow area, was confirmed on histopathology.

• *Final diagnosis: Large, locally invasive, poorly differentiated squamous cell carcinoma.* The patient was referred to the oncologist, to be followed up in the outpatient setting. He elected to receive definitive therapy at an out-of-state tertiary cancer center. He was discharged on 4/2/09. Before discharge from the hospital, nuclear medicine positron emission tomography (PET) scan showed no evidence for lymphoma, but confirmed the known increased activity in the left elbow.

Postscript

Over the next several months, the patient received chemotherapy and radiation therapy at the outside cancer center. He was readmitted to the local hospital several times through May, 2010, primarily for complications of treatment of his cancer. Those complications included weakness and anemia requiring transfusion and, especially, left arm pain and swelling after left axillary lymph node dissection in late 2009. A biopsy of an axillary lymph node on 11/24/09 before the dissection had confirmed "metastatic sarcomatoid carcinoma with extensive necrosis." He died less than 2 years after his initial cancer diagnosis.

Case discussion

This was the case of an 82-year-old Caucasian male with persistent or non-healing cellulitis and abscess in the left elbow, even after several months of seemingly appropriate antimicrobial therapy. The organism cultured from the site on several occasions was methicillin-sensitive *Staph. aureus*, that should have responded to the given antibiotics. Examination of the patient suggested lymphatic obstruction, especially with left axillary lymphadenopathy and persistent edema of the arm. A tissue biopsy was undertaken and locally invasive squamous cell cancer was diagnosed. Radiologic studies, including a PET scan, did not confirm metastasis or lymphoma at the time of the initial cancer diagnosis. The patient had complications of cancer therapy (anemia, weakness, arm pain and swelling), especially after axillary node dissection. He died less than 2 years after the initial cancer diagnosis.

Lessons learned from this case

- The most important lesson (for physicians) from this case is timely referral of a patient who is not responding as expected to seemingly appropriate therapy.
- It is, however, unclear whether his cancer would have had a different outcome had he been referred earlier, and an earlier diagnosis of malignancy made.
- The patient was also generally "stoic" and required a lot of encouragement from his family even to keep already scheduled appontments. He initially shrugged off the "bump" in the elbow for several months before the first visit to the doctor.
- It was still unclear whether this squamous cell cancer was in any way related to the alleged non-Hodgkin's lymphoma that was previously diagnosed some 4–5 years before the first patient encounter.
- Finally, the infectious diseases consultant must evaluate the patient completely (detailed history and complete physical examination) and keep in mind that the patient's illness or problem may be non-infectious, even if that was the thinking of the referring practitioner.

Further reading

Falagas ME, Vergidis PI. Narrative review: diseases that masquerade as infectious cellulitis. Annals of Internal Medicine 2005; 142: 47–55. Available at: http://keck.usc.edu/en/Education/Academic_Department_and_Divisions/Department_of_Medicine/Education_and_Training/Internal_Medicine_Residency/Resources/Articles/~/media/Docs/Departments/Medicine/Chief%20Resident/Articles/ID/Essential/Cellulitis.pdf, accessed February 25, 2016.

CHAPTER 4

Diseases Acquired Through Close Contact with Animals

Case 4.1   A 10-month-old child with "worms"

A 10-month-old male child was brought by the mother to the office on 12/21/05. The mother had noted something in the stool that moved and looked like worms. She had been referred by a veterinarian she had sought help from. The child looked healthy, and not ill.

In October, 2005, the child had received two courses of mebendazole from the primary care physician within 2 weeks, but with no response. The mother continued to see these worms intermittently.

The review of systems showed a child with no fever or diarrhea (stool was generally normal). He had some itching around the buttocks when the diaper was removed. He had no rash, however, on the buttocks or anus area, and he was healthy looking and very comfortable.

Epidemiologic history: the mother and the child's 2-year-old brother were healthy. They had all spent time after Hurricane Katrina (which occurred on 8/29/05 on the Gulf Coast) living with relatives (in crowded facilities) in September and October, 2005. There were several dogs in the overcrowded abode and the environment was shared with many other displaced relatives and friends. Dogs were said to have defecated on the floor. At least two of these dogs were indoors, while others were outdoor animals. Finally by December, 2005 they had moved on to another home.

The past medical history showed nothing significant. The child was circumcised soon after birth. His immunization status was unknown, but presumed up to date.

Family history was significant for a mother aged 22 and a father aged 23 years. The child had a 2-year-old brother. Nothing else was significant; other family members were said to be healthy.

On examination, he appeared very healthy, comfortable, and in no acute distress. His vital signs showed a pulse of 120, respiratory rate (RR) 36, height 28″, weight 25 pounds, and temperature 98.6 °F. He had normal head, neck, and chest (heart and lung) exam. The abdomen, extremities, and skin were unremarkable.

Cases in Clinical Infectious Disease Practice: Obtaining a Good History from the Patient Remains the Cornerstone of an Accurate Clinical Diagnosis: Lessons Learned in Many Years of Clinical Practice,
First Edition. Okechukwu Ekenna.
© 2016 John Wiley & Sons, Inc. Published 2016 by John Wiley & Sons, Inc.

The skin, including the buttocks, perineal and perianal exam, was without rash or other visible pathology.

Stool, including scotch tape collected specimen and a second stool sample, was examined for parasites, with negative results for ova and parasites.

Follow-up visits

The patient was seen again 1 month later, on 1/23/06, in the office. This time, the mother brought with her a transparent bag with stool material she had collected 2 hours earlier that she thought were wriggling worms. The material she brought was examined (visually inspected), and there was some observed movement of rice-like (estimated at 2–3 mm size), whitish-creamy colored material that could be a proglottid. This material was immediately sent to the laboratory for identification, with instructions to forward it to the state health department laboratory as well.

The physical exam of the patient found him again to be very comfortable, with no signs of irritation around the buttocks, except some diaper irritation in the groin area.

Laboratory investigations

Over the next 2 months, between January and March, 2006, additional specimens were sent to the laboratory for parasite identification. We stayed in touch with the mother. We also reviewed previous reports of studies done elsewhere. The specimens from our institution were forwarded to the state health department laboratory for confirmation, or to rule out the presence of parasites. These stool specimens were from the following dates: 12/23/05, 1/23/06, and 2/3/06. The stool specimens processed elsewhere prior to the office visit were on the following dates: 11/04/05, 11/23/05, 12/14/05, and 12/19/05. All of these were reported negative for ova or parasites by these laboratories. There was no eosinophilia noted.

The complete blood count (CBC) on 3/25/06 showed the following: white blood cell (WBC) count was 11,400/μL, hemoglobin and hematocrit (H/H) 12.5/37.3, respectively, and platelet count 513,000/μL; differential count showed 15% polymorphs, 75% lymphocytes, 3% atypical lymphocytes, 6% monocytes, and 1% eosinophils.

On 3/13/06, one stool specimen was brought by the mother to the laboratory that had more worms than previous specimens. Fig. 4.1a shows a photograph of the material in formalin, taken at the time. This material was again forwarded to the state health department, and subsequently to the Centers for Disease Control and Prevention (CDC) in Atlanta. We had confirmed at this time that this was some type of tapeworm, but which type?

• *What is your diagnosis at this time and how would you treat this patient?*

Figure 4.1a Proglottid of tapeworm obtained from the 13-month-old Caucasian male (March, 2006). (*See insert for color information*)

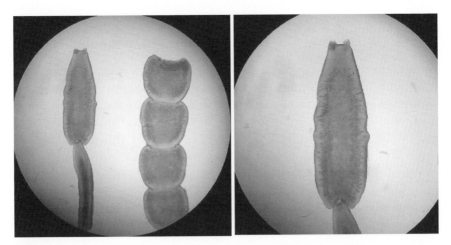

Figure 4.1b These digital images depict tapeworm proglottids from our patient: the mature proglottid on the right shows lateral genital pores and "rice-grain" shape, while the immature proglottids are broader in appearance. Courtesy of Henry S. Bishop, CDC, Atlanta. (*See insert for color information*)

It took a few weeks of discussion with the CDC to confirm the diagnosis and discuss possible treatment in this very young child. We also needed to update the primary care physician about the issues involved before any formal treatment could be given.

Additional photomicrographs were sent to us by the CDC from the patient material (stool and adult tapeworm) forwarded to them (Figs 4.1b, 4.1c).

- *Final diagnosis: Dipylidium caninum (dog flea tapeworm).*

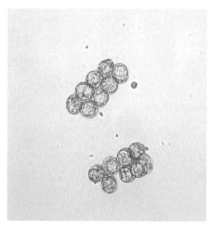 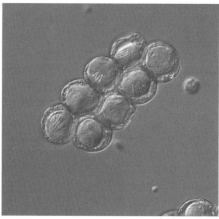

Figure 4.1c Eggs in a cluster: image captured by use of digital image correlation (DIC). Individual eggs are about 35–40 µm (31–50 × 27–48 µm), with an onchosphere that has six hooklets visible. Courtesy of Henry S. Bishop, CDC, Atlanta.

Case discussion

This case has many facets to it. First, it was difficult to make a diagnosis in a timely manner. The mother was persistent, while the labs were unable to identify the tapeworm or the eggs. The proglottids (which are broken-up parts of the adult worm) did not survive for very long after separation from the adult worm, and often were not recognized as such when they arrived at the state department labs. It took several discussions with the labs, insistence that the mother was correct in her observation, and possibly luck (because we finally received a good enough specimen with a long portion of an adult worm which better defined the nature of what we were looking at) to come to a definite diagnosis. Both the character of the proglottid and the embryonated eggs were necessary to come to a definitive identification of the worm.

The other issue of importance and difficulty was what to do after a diagnosis was made. The child was not particularly ill from this parasite, except for some itching, but the mother was very concerned. Besides, there was no recommended therapy for this type of infestation for children younger than 4 years. The child was 10 months when initially seen, and 13 months when a final diagnosis was made.

Discussions with staff at the CDC and review of available literature confirmed that there was little published material on treatment in children less than 4 years with praziquantel. However, this drug has been around for many years, and has been used worldwide in many young children and small animals to treat a variety of parasitic infections without any serious adverse effects. A single dose would be required, without the need to treat the rest of the family. These discussions and review led to the delay in the diagnosis and treatment of this patient.

On 4/12/06, the mother was called to the office. She came with the two children, the 2.5-year-old brother and the now 14-month-old patient. Both were well and seemingly asymptomatic.

After full informed consent, the mother consented to treatment with praziquantel, based on the most recent Medical Letter [1] on treatment of parasitic infections. The patient was prescribed a one-time dose of 10 mg/kg of praziquantel. This treatment was reviewed with the pharmacist (who compounded the medication) and the primary care physician.

He received his treatment, and has had no more problems related to this parasite since. The last contact with the mother was 3 years after the treatment (in 2009), and all was well. No other family member required or needed treatment.

Additional comments on this case

One of the misconceptions about this parasite is that the infection is caused by ingesting the worm or parasite. That is not the case. It is the dog flea that contains the parasite that is ingested. The life cycle of this dog flea tapeworm is depicted in Fig. 4.1d. *Dipylidium caninum* (also called the double-pored dog tapeworm)

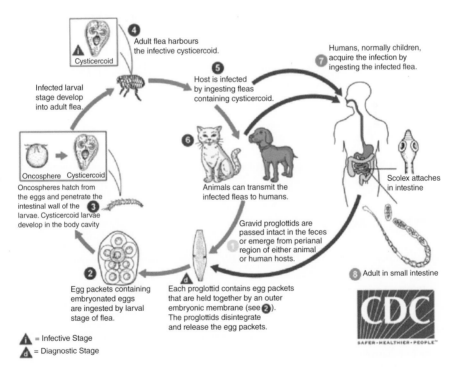

Figure 4.1d Life cycle of *Dipylidium caninum* (the double-pored dog tapeworm) which mainly infects dogs and cats, but is occasionally found in humans. (The expanded legend can be accessed at www.cdc.gov/dpdx/dipylidium/index.html.)

mainly infects dogs and cats, but very occasionally is found in humans. Such humans are usually very young children who inadvertently ingest the dog or cat flea as they put other things into their mouths (as is typical for that age). This explains why this is not a disease of adults unless under pathologic conditions in a severely mentally retarded individual in the appropriate circumstances.

It is presumed that as a 7–8-month-old baby, crawling around in a contaminated and crowded environment with infected dogs, likely infested with fleas, this patient may well have ingested some of the flea that started a life cycle in his gut, and led to this infestation. His older brother and the adults in the same environment did not have this problem because they were less likely to ingest fleas. This is typically a disease of toddlers.

The mother's persistence led us to a final diagnosis, and relief for her and her family.

Lessons learned from this case

- Do not discount the patient's history (in this case that of a very observant mother).
- Persistence on the part of the mother (and us) eventually paid off.
- Work with the laboratory, but do not allow them to talk you out of obvious clinical observation (the proglottid was seen to move, and eventually a larger tapeworm was finally brought in).
- We were able to work with the CDC parasitology division to resolve this issue. The photomicrographs are from the patient's specimen processed by the CDC.

Reference

1 Medical Letter: drugs for parasitic infections (August 2004). Available at: www.medicalletter .org, accessed February 26, 2016.

Further reading

Centers for Disease Control and Prevention. Life cycle of *Dipylidium caninum*. Available at: www .dpd.cdc.gov/dpdx/HTML/Dipylidium.htm, accessed February 26, 2016 (see details in the CDC's Parasite Image Library, including the figure legend).

Diseases from dogs. Available at: www.cdc.gov/healthypets/pets/dogs.html, accessed February 26, 2016.

Fact Sheet for the general public: Dog and cat flea tapeworm. Available at: www.cdc.gov/ parasites/dipylidium/index.html, accessed February 26, 2016.

Kotton CN. Zoonosis from dogs. Available at: www.uptodate.com/online/content/topic.do? topicKey=oth_bact/4546&view=print, accessed February 26, 2016.

Case 4.2   Infectious complication of cat scratch

A 32-year-old Caucasian female was referred in late January, 2005, for skin abscesses and rash that would not heal for about 7 weeks.

On 12/2/04, she had taken her sick cat to the veterinarian for evaluation. The cat was ill with sores and had shortness of breath. On the way to the veterinarian's office, this cat was clinging tightly to her with its claws, and in the process had scratched her in the neck, shoulder, abdomen, and flank areas. The scratched areas remained sore and bruised, in spite of various courses of antibiotics received over a 7-week period. The antibiotics included cefdinir, cephalexin, and, most recently, moxifloxacin (last dose of a 14-day regimen completed 1 day prior to the office visit).

She had six cats, three of which had died within the last 2 months, between 3 and 4 weeks apart. The last ill cat was "put down" by the veterinarian because of sores. Her three outdoor cats were healthy, as were her two dogs.

The review of systems showed her to have no fever. The multiple skin sores continued to drain, in spite of local treatment with soap and application of topical triple antibiotic ointment, and the use of systemic antibiotics. Some of the sores were now showing some tunneling, and they were tender, painful, and burning.

The past medical history was significant for migraine headaches for years; arthroscopic right knee surgery for a patella problem several years previously; and two previous C-sections 11 and 13 years earlier. Four months earlier she had a knee abscess with bursitis for which she was given gatifloxacin, which was now healed. She had tonsillectomy and adenoidectomy at age 16. Her immunizations were up to date.

Family and social history was significant for the following: she lived in a rural southern Mississippi community, was divorced, had two jobs as a hairdresser and waitress, and was financially distressed. She rarely smoked cigarettes or drank alcohol.

She was allergic to penicillin but had been able to tolerate cephalexin recently.

On examination, she was alert and oriented ×3. Her vital signs showed that she was afebrile, had a blood pressure (BP) of 108/76, RR 18, heart rate (HR) 80; height 5′3″, weight 205 pounds. The head and neck, heart, lung, abdominal, upper and lower extremity exams were unremarkable. The neurologic exam was normal.

However, her skin showed tiny, tunneling, and superficial subcutaneous abscesses in the right upper arm, right shoulder, left lateral flank, and right mid abdomen, northwest of the umbilicus. She also had multiple subcutaneous nodules noted in the right arm and shoulder areas.

Figures 4.2a and 4.2b are samples of photographs of the skin taken during the first office visit on 1/27/05.

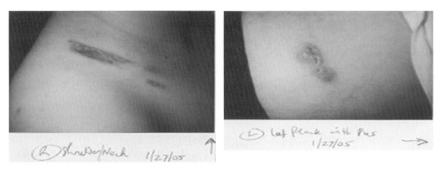

Figure 4.2a Right shoulder/neck and left lateral flank areas with tiny tunneling abscesses (reproduced with permission). (*See insert for color information*)

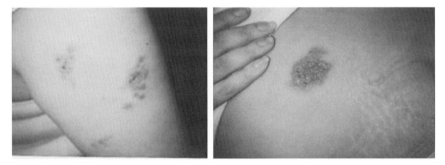

Figure 4.2b Right upper arm and abdomen (above and right of the umbilicus): tiny tunneling abscesses (reproduced with permission).

Thick creamy pus was obtained from the left lateral flank area after the scabbed abscess was unroofed. The specimen was sent for gram stain and cultures, to include fungal and mycobacterial stains and cultures.

• *What are the differential diagnoses to consider at this time?*

Antimicrobial treatment was initially withheld until follow-up visit in a few days unless culture results were available earlier (the patient had just completed several courses of antibiotic treatment). She was seen 5 days later in the office. Examination was unchanged, except for more prominence of the subcutaneous nodules in the right shoulder region. The gram, KOH, and acid-fast bacillus (AFB) stains, as well as cultures, were negative; the differential diagnoses thus included mycobacterial or fungal infection. To cover both presumptively would require at least two expensive drugs, so we opted to wait a while.

Follow-up and response to treatment

The patient was seen several times in the office. She returned about 4 weeks after the initial cultures were obtained. A tiny mold growth was reported by the laboratory. She was started on presumptive treatment with oral itraconazole, pending final culture results.

She came back 19 days later on 3/21/05 with new abscesses in the right lower abdomen, left groin, and upper thigh areas (where she had shaved) that turned out to be due to MRSA. She was hospitalized for 3 days (3/25–3/28/05) for abscess drainage. In addition to the itraconazole for the fungus, she received trimethoprim/sulfamethoxazole to cover the MRSA. Three weeks later, there were no palpable nodules noted. Her skin lesions began to dry up within weeks.

We encouraged her to complete at least 3 months of therapy with itraconazole.

The culture was confirmed positive for *Sporothrix schenkii* after 4 weeks in culture, with final confirmation by thermal dimorphism, and tease mount culture characteristics.

Case discussion

This patient developed abscesses after her sick cat clawed her skin in many areas, including the right upper arm and shoulder, right abdomen, and left flank. The abscesses did not heal after several weeks of antibiotic therapy so she was referred to see us. There was tunneling of many of the abscesses. The initial gram stain, KOH, and AFB stains were negative, and the bacterial cultures were negative after a few days. At that time, the differential causes considered included subcutaneous fungi or mycobacteria. It took another 3 weeks to notice some growth, and by 4 weeks, she was started on itraconazole to cover sporotrichosis.

We discussed the option of putting her on two drugs, to cover both *Mycobacterium marinum* and sporotrichosis, but the expense of the drugs caused us to delay that decision. Infections due to *M. marinum* and sporotrichosis can have similar clinical appearance in presentation, although the epidemiology is quite different. Nodules along the lymphatics are common for both.

The patient was able to obtain some samples of other antimicrobials which she took in the interim (clarithromycin), until she was started on itraconazole. She was already on itraconazole when she developed new abscesses in the right lower abdomen and left groin and upper thigh, likely related to shaving in the areas. The MRSA cultured was unrelated to the sporotrichosis. The skin abscesses healed over a period of weeks, with residual superficial scarring when seen 11 weeks after the initial office visit.

Key to the microbiologic diagnosis was the growth and cultural characteristic of *Sporothrix schenckii* as a thermally dimorphic fungus (like *Histoplasma*, *Blastomyces*, and *Coccidioidomyces*). Brain heart infusion (BHI) agar with blood incubated at 37 °F produced a yeast-like culture, while culture on Sabouraud dextrose agar (SDA) at 25–30 °C (room temperature equivalent) showed moldy growth characteristics. Finally, tease mount of the fungus growth confirmed septate hyphae bearing unicellular conidia on denticles typical for *Sporothrix schenkii* (see Figs 4.2c–4.2e).

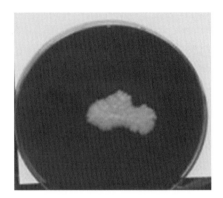

Figure 4.2c Brain heart infusion (BHI) agar (with blood), 4-week growth of yeast-like colony.

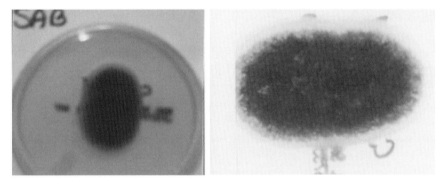

Figure 4.2d Sabouraud dextrose agar (SDA), 3–4-week culture. Left, bottom of agar; right: aerial mold growth.

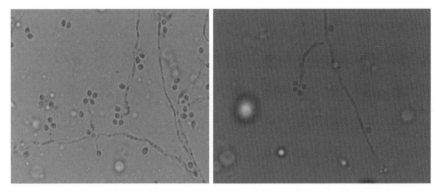

Figure 4.2e Tease mount of culture grown on potato flake agar showing septate hyphae bearing unicellular conidia on denticles, typical for *Sporothrix schenkii* (lactophenol cotton blue stain).

The clinical features of sporotrichosis include lymphocutaneous, pulmonary (a more serious disease), osteoarticular, and meningeal factors; other forms include disseminated and other localized visceral infections.

Infection is usually acquired through exposure to soil, wood (e.g. puncture through splint), or plant material. People involved in landscaping, farming or gardening may have a particular professional exposure to this organism [3].

Outbreaks in cats have been reported, with zoonotic infections affecting mainly women and children (those more likely to come in contact with cats) [3,1].

In our patient, the organism took nearly 3 weeks to be identified as a fungus. In many cases, however, *S. schenckii* is considered to have rapid growth (within 1–2 weeks, typically), even if it may take longer to reach a final identification with thermal dimorphism tests and other characteristics. It is possible that other previous treatments, co-infection with other agents, or low inoculum of the cultured specimen may have influenced the growth rate of the *S. schenckii* in this case. Incidentally, 2–3 months earlier we saw another patient with a thumb abscess that was caused by a wood splinter. *S. schenckii* was cultured within 10 days in that patient.

Our patient had the lymphocutaneous form of sporotrichosis, the most common form of the disease seen in clinical practice. It is likely she acquired the infection from cat sores, contact with cat mucous membrane, or other contamination (or inoculation) of the clawed skin with the organism from the infected cat.

Treatment guidelines have been addressed in the Infectious Diseases Society of America (IDSA) guidelines for the management of patients with sporotrichosis noted below [2].

Postscript
I spoke with the patient in 2015, 10 years after her infection. She continues to do well, as expected. She has not reported any more infected cats.

Lessons learned from this case
- Sporotrichosis is a subacute or chronic infection that most commonly involves the skin and subcutaneous tissue, and then the lymphatics.
- It is easily confused with *M. marinum* infections, because of clinical similarity and the presence of subcutaneous nodules in the lymphocutaneous variety.
- As with *M. marinum*, the skin infection is often indolent and persistent. However, the epidemiology is different.
- Routine bacterial cultures are negative. That fact should lead to further questions.
- Appropriate fungal cultures must be done otherwise the diagnosis will be missed.

- Sporotrichosis can be a lethal disease in cats, as with several of this patient's cats. The dead cats (all indoor cats) probably had disseminated disease, with generalized sores, mucosal involvement, and pulmonary disease.
- Skin abscesses that do not respond to usual antibacterial treatment should be further assessed clinically, and with additional fungal and mycobacterial cultures.
- It is also important to culture an abscess before treatment is started, as a negative culture initially would have alerted the physician to look for other possible causes.

References

1 Kauffman CA. Basic biology and epidemiology of sporotrichosis. Available at: www.uptodate.com/online/content/topic.do?topicKey=fung_inf/15174&view=print, accessed February 26, 2016.

2 Barros MBL, Schubach AO, Schubach TMP, Wanke B, Lambert-Passos SR. An epidemic of sporotrichosis in Rio de Janeiro, Brazil: epidemiological aspects of a series of cases. Epidemiology & Infection 2008; 136: 1192–6. Available at: www.ncbi.nlm.nih.gov/pmc/articles/PMC2870916/pdf/S0950268807009727a.pdf, accessed February 26, 2016.

3 Kauffman CA, Bustamante B, Chapman SW, Pappas PG. Clinical practice guidelines for the management of sporotrichosis: 2007 Update by the Infectious Diseases Society of America. Clinical Infectious Diseases 2007; 45(10): 1255–65.

Further reading

Kauffman CA. Treatment of sporotrichosis. Available at: www.uptodate.com/online/content/topic.do?topicKey=fung_inf/14834&view=print, accessed February 26, 2016.

Kauffman CA. Clinical features and diagnosis of sporotrichosis. Available at: www.uptodate.com/online/content/topic.do?topicKey=fung_inf/7293&view=print, accessed February 26, 2016.

Case 4.3   A young man with acute left arm lymphadenopathy

A 29-year-old Caucasian male was admitted in January, 2009 for lymphadenopathy. Two weeks earlier, he had noted a pimple-like nodule in the left arm above the elbow, with no visible pus or drainage. Several days later, he reported some chills and low-grade temperature. Four days prior to admission, clindamycin was prescribed by his primary care physician. He did not improve. He was therefore referred to the orthopedic surgeon for further evaluation, including consideration for biopsy of the lymph node. He was admitted for inpatient evaluation.

The next day, on 1/14/09, magnetic resonance imaging (MRI) study showed two enhancing nodules above and medial to the left elbow, with the largest one measuring about 2 cm in size. Ultrasound (US)-guided needle biopsy was done, with the specimen sent for histopathology and cultures.

Review of systems was significant for the following: he had a palpable nodule in the area of the left arm above the medial elbow. He had also noted a pustule in the left hand between the fourth and fifth fingers near the web of the interdigital space, which had been present for 3 weeks or longer. The lymph nodes in the left arm had been present probably for 10–14 days. He had low-grade temperature, associated with chills, mostly in the evenings, and intermittently. There was a localized erythema over the site of the swelling in the left arm. Relevant epidemiologic history included the fact that his wife and two children (8-month-old son and 4-year-old daughter) were healthy. He had not gone fishing or hunting in the past year; nor had he been involved in gardening or handling of roses or other plants. This patient's job was in an office where he worked as an engineering designer.

He had no prior serious medical conditions or surgeries, and did not drink alcohol or smoke. He had no known drug allergies. Prior to admission, he had been started on clindamycin. After admission to the hospital, he was started on vancomycin and clarithromycin.

Infectious disease consultation was placed 2 days after admission to help with the diagnosis.

On examination, he was alert and oriented, and in no acute distress. Vital signs were as follows: BP 135/80, RR 16, HR 83, temperature 97.9 °F; height 5'10", weight 204 pounds. There were no significant findings in the examination of the head, neck, heart, lung, and abdomen. Lymph nodes were not palpably enlarged in the neck, axilla, or groin. However, nodules were palpable in the left upper arm, medially above the elbow, at the site of the recent biopsy the day before, on 1/14/09. The area was raised and nodular, but not very tender. No other significant adenopathy was noted in the epitrochlear region or in the left axilla. The left hand showed a minimally erythematous, old, chronic, and healing pustule in the web of the fourth and fifth fingers. The skin showed a diffuse, erythematous, blanching, partially petechial type, fine rash, which did not itch. This rash involved the torso and the upper and lower extremities, but not the palms or soles. Other examinations, of the musculoskeletal system, external genitals, and neurologic system, were unremarkable.

A tentative diagnosis was made on the basis of the clinical findings and history. Serologic tests were ordered and treatment was started presumptively. The patient was discharged after the ID consult to outpatient follow-up. The rash was thought to be an adverse reaction to one of the antibiotics – vancomycin, clarithromycin, or clindamycin – previously received.

• *What are the differential diagnoses that come to mind for this patient?*

The patient was seen 8 days later (1/23/09) in the office for a scheduled follow-up visit. The left arm lymph node bulge was still present, with no discharge or drainage noted. The rash was resolved. We had stopped the clarithromycin and vancomycin, and placed him on doxycycline on 1/15/09. His physical exam was unremarkable, except for the bulging area above the left medial elbow. There was minimal tenderness to palpation of this area that appeared to have more than one nodule, and was several centimeters across (Fig. 4.3a).

• *What is your diagnosis at this time? What serologic test do you think was ordered in the hospital on 1/15/09?*

The serologic test ordered during the recent hospital admission was incorrectly transcribed, and so was not available during the outpatient follow-up visit. It was reordered from the office. The fine needle aspiration lymph node biopsy obtained through ultrasound guidance on 1/14/09 was described as "reactive," with no evidence of lymphoma. The histopathology showed a mixed acute and chronic inflammatory infiltrate with microabscess formation. Routine culture was negative while special stains and other cultures (AFB and fungus) were pending.

The patient was continued on doxycycline, and asked to return for follow-up in 2 weeks.

On return to the office on 2/6/09, he was much better. The swelling in the left arm had shrunk more than 90%. The patient confirmed that about 4 days after the office visit of 1/23/09, he had noted some serous drainage at the site of the bulge and needle biopsy. The bulge had collapsed, with only residual tiny palpable nodules felt at the old site on palpation. He had no fever, and no erythema or drainage. Figure 4.3b was taken 22 days after doxycycline was started.

During this visit, the serologic test result was available, confirming the clinical diagnosis made during the initial consult of 1/15/09.

• *Final diagnosis: Cat scratch disease (CSD), due to Bartonella henselae.*

Case discussion and postscript

Crucial epidemiologic information obtained during the initial ID consultation was withheld in order not to give away the diagnosis prematurely. This patient

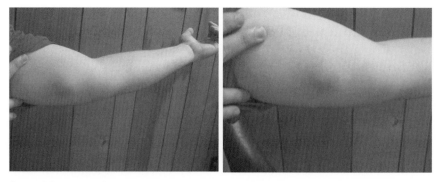

Figure 4.3a Photograph of the left arm showing bulging mass above the medial elbow, 1/23/09 (reproduced with permission).

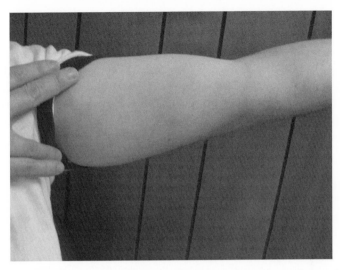

Figure 4.3b Resolved lymphadenopathy of the medial left elbow. Photo taken on 2/6/09, 22 days after doxycycline was started (reproduced with permission).

and the family had taken in a 5-month-old kitten about 1 month before the hospital admission. They did not know where the kitten came from originally, but it was said to be healthy, with no skin lesions. He could also not recall any cat scratch or bites. However, he had a nodular pustule in the left hand in the web of the fourth and fifth fingers that was healing at the time he was admitted to the hospital. The bulge in the left medial elbow was consistent with lymphadenopathy, as was confirmed by the MRI, and the histopathology of the fine needle biopsy.

Serology by immunofluorescence assay (IFA) for CSD (*Bartonella henselae*) on 1/23/09 was positive for both IgG and IgM antibodies: 1:256 for IgG (normal range <1:64) and 1:80 for IgM (normal range <1:20).

The lesion in the left hand (pustular nodule) was healed at the time of the 2/6/09 office visit, but had been present for 3 weeks at the time of the initial consultation on 1/15/09. The lymphadenopathy was noted 1 month after acquisition of the cat. We think the infection was acquired through contact with the then 4-month-old "healthy" kitten, in spite of no documented cat scratch or bite.

Lessons learned from this case
- This case underlines the importance of obtaining a thorough history from the patient in making an accurate diagnosis.
- The lymph node biopsy was done because of a suspicion of cancer but only because an incomplete and inadequate history was taken in this otherwise healthy person.
- The lymph node biopsy was unhelpful, but it ruled out a cancer diagnosis.

- The recent acquisition of a kitten in the house, and the finding of a healing pustule in the left hand, in the lymphatic drainage area of the left elbow lymph node, suggested the likelihood of CSD in this patient.
- Serologic confirmation was delayed initially because of a laboratory test entry error.
- By the time of the follow-up office visit, the adenopathy was resolved.
- The unrelated adverse drug reaction (rash) the patient had did not distract from the diagnosis of CSD.

Further reading

Carithers HA. CSD: an overview based on a study of 1200 patients. American Journal of Diseases of Children 1985; 139(11): 1124–33. Available at: www.researchgate.net/publication/19101353_Carithers_HA_Cat-scratch_disease_An_overview_based_on_a_study_of_1200_patients_Am_J_Dis_Child_39_1124-1133, accessed February 26, 2016.

Spach DH, Kaplan SL. Microbiology, epidemiology, clinical manifestations, and diagnosis of cat scratch disease. Available at: www.uptodate.com/online/content/topic.do?topicKey=oth_bact/5634&selectedTitle=1%7E44&source=search_result, accessed February 26, 2016.

CHAPTER 5

Travel-associated Blood-borne Parasitic Infection

Case 5.1 A 15-year-old adolescent male with fever and thrombocytopenia

A 15-year-old adolescent male from El Salvador was admitted in late October, 1998, with a 6-day history of fever and chills.

He had traveled with his parents to visit relatives in South Mississippi. The trip took 6 days by road from El Salvador (capital city), via Guatemala, through Mexico, Arizona, through New Mexico, Texas, and finally to South Mississippi.

The past medical history showed no previous serious underlying diseases, other than a remote history of previous amoeba in the stool.

He had two siblings at home in El Salvador, aged 18 and 13 years; both were said to be healthy. Both parents and the relatives in South Mississippi were healthy. The patient did not drink or smoke. He had no known drug allergies.

The review of systems showed that he had arrived in South Mississippi 15 days earlier, 13 days prior to admission (PTA), following the 6-day trip by road. High fever with chills had started 6 days PTA (8 days before the infectious diseases consult; 1 week after arrival in South Mississippi). Chills and rigors occurred on average about 1–2 times per day. Associated headache occurred usually with the fever, at which time he would have occasional sweating and might be dizzy. In between the fever and chills, he felt well.

He also took over-the-counter cough medicine for dry cough. He had no rash and denied other significant symptoms (no nausea, vomiting, diarrhea, body or joint pains).

The epidemiologic history was significant for the following: the patient was well when they left El Salvador for the trip. He was a city dweller (not a rural kid). There was no significant known family history of unusual illnesses (both parents were well). In South Mississippi, the children in the family they were visiting (ages 10, 3 and 1) were all healthy. During their 6-day trip, they slept

Cases in Clinical Infectious Disease Practice: Obtaining a Good History from the Patient Remains the Cornerstone of an Accurate Clinical Diagnosis: Lessons Learned in Many Years of Clinical Practice,
First Edition. Okechukwu Ekenna.
© 2016 John Wiley & Sons, Inc. Published 2016 by John Wiley & Sons, Inc.

in hotels along the way. Nothing special or unusual happened during the trip, according to the patient and his parents. There were no unusual contacts with exotic or other animals.

One day after admission, hematology-oncology consultation was sought to evaluate the fever in this young patient, who also had pancytopenia and severe thrombocytopenia.

Additional consultation was requested with infectious diseases the next day (2 days after admission), to address the fever of unknown origin and thrombocytopenia.

On examination, he was found to be a pleasant, mildly ill, asthenic young man in no acute distress. He was fully alert and oriented.

Vital signs were as follows: blood pressure (BP) 114/44, respiratory rate (RR) 24, heart rate (HR) 87 beats per minute, maximum temperature 104.4 °F; height 5'11" and weight 118 pounds. The head and neck exam showed minimal conjunctival injection, but otherwise was unremarkable. The heart, lung, abdominal, external genitalia (uncircumcised male), extremities, and neurologic examinations were normal. Lymph nodes were not palpably enlarged in the neck, axilla, or groins. The skin showed only mild facial acne, but otherwise no rash.

• *At this stage what differential diagnoses would you consider?*

Laboratory and other investigations

The chest x-ray (PA/Lat/oblique) showed a "benign coin nodule" in the left upper lung. Abdominal ultrasound showed "debris" in the gallbladder. Skin tests for purified protein derivative (PPD), *Candida*, and histoplasmin were read as negative. The bone marrow (BM) biopsy and aspirate (from the iliac crest) done on 10/29/98 by the hematologist showed reactive hyperplastic marrow, with no malignancy identified. There were scattered plasma cells, approximating 4%; the myeloid:erythroid ratio was 3:1. Additional BM staining results were reported several days later on 11/4/98.

Serologic tests for Epstein–Barr virus (EBV) were positive for viral capsid antigen (VCA) IgG at 1:1280 and nuclear antigen (EBNA), but negative for all other parameters (VCA IgM, and early antigen [EA] IgG). Blood cultures ×2 on admission were negative and HIV serology was negative. Sedimentation rate was mildly elevated at 28 mm/hour. Admission complete blood count (CBC) showed a white blood cell (WBC) count of 4.2, hemoglobin and hematocrit (H/H) 11.9 and 34.3, respectively, mean corpuscular volume (MCV) 76.9, and platelets 34,000 (31,000 on 10/29/98). The differential blood count was as follows: 34% polymorphs, 29% band forms, 22% lymphocytes, 5% atypical lymphs, 10% monocytes, and 0.3% reticulocytes. Chemistries showed a sodium level of 133, potassium 3.6, glucose 111 mg/dL, and BUN/creatinine of 11 and 0.9, respectively. Lactate dehydrogenase (LDH) was 778 (normal range: 313–618 IU/L);

serum glutamic oxaloacetic transaminase (SGOT) was 92, serum glutamic pyruvic transaminase (SGPT) 54; alkaline phosphatase (AP) level was 260, total bilirubin 1.3, albumin 3.1 g/dL (2.5 on 10/29/98).

Normal or negative labs included urinalysis and culture, stool studies for routine culture, ova, cysts and parasites, and *C. difficile* toxin assay. Acute hepatitis A, B, and C serology, amylase and lipase levels, monospot test, and antinuclear antibody (ANA) tests were negative or normal. Iron (Fe) concentration was low (35 μg/mL), total iron-binding concentration (TIBC) and ferritin levels were normal at 284 μg/dL and 216 ng/mL, respectively.

- *Would you change your differential diagnoses based on these laboratory test results?*

Hospital course

The infectious diseases (ID) consultant discussed the differential diagnoses, and recommended a diagnostic procedure based on the epidemiology, symptomatology, and physical examination findings: 8 days of fever with chills, no localizing signs, no lymphadenopathy, no eosinophilia, but significant thrombocytopenia. Fig. 5.1a is a graphic depiction of the vital signs recorded between 10/28/98 and 11/1/98.

- *What diagnostic procedure do you think was recommended by the ID consultant?*

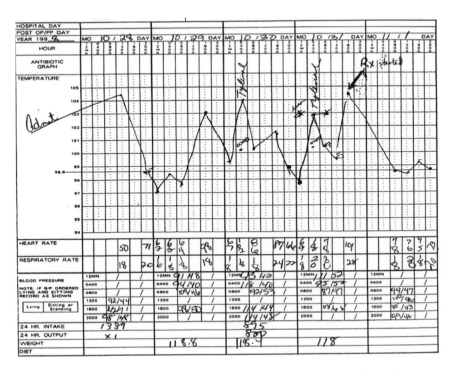

Figure 5.1a Recorded temperature and other vital signs between 10/28 and 11/1/98.

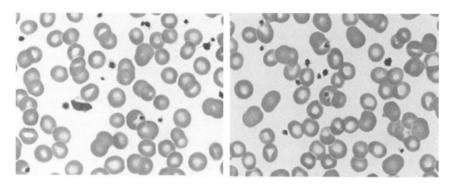

Figure 5.1b Thin smear of blood (Giemsa stain) for malaria, showing the "banana-shaped" gametocyte, typical for *P. falciparum*, on the left slide and "ring" forms noted on both smears. Smear was made on 10/31/98.

Follow-up of patient

Specific therapy was started on 10/31/98, and the patient was discharged to out-patient follow-up 2 days later on 11/2/98. He was hospitalized from 10/28 to 11/2/98. On 11/6/98, WBC was 6.2, H/H 11.6/35.1, MCV 78.7, and platelet count was normal at 397,000. Differential counts were as follows: 36.6P, 49.4L, 12.5M, and 0.9% eosinophils.

The result of the BM biopsy special stains (done on 10/30/98) was reported on 11/4/98, officially confirming the previously made diagnosis (Fig. 5.1b).

At the outpatient follow-up visit on 11/9/98, 1 week after discharge from hospital, the patient was healthy looking, with resolution of fever and other symptoms. Thrombocytopenia was also resolved (397,000/μL on 11/6/98).

- *Final diagnosis: Plasmodium falciparum malaria infection.*

Comments

Among other findings in the bone marrow biopsy were the following: decreased stainable iron; no metastatic tumor or hematopoietic malignancy, but rare malarial forms were present in the mature red blood cells (RBC) seen in the aspirate smears.

Fever resolved promptly after starting chloroquine phosphate on 10/31/98, for a 3-day treatment. This regimen was chosen because there was no chloroquine resistance in Central America at the time. Follow-up thin smear for malaria performed on 11/6/98 showed no "ring" forms but there were still rare macrogametocytes ("banana" forms) persisting. However, the macrogametocyes do not cause the febrile symptoms of malaria, do not respond to chloroquine therapy, and would disappear over time. Their role in any subsequent infection would require their being taken up during a mosquito bite meal for processing in the insect life cycle.

Case 5.2   A 23-year-old Caucasian male with fever and chills

A 23-year-old Caucasian male was admitted in May, 1999 with fever and chills. Six days prior to admission, he complained of nausea and vomiting, then fever with chills, several times a day. He would take ibuprofen, which caused him to sweat. He was fatigued. He took over-the-counter "flu" medicines without effect. Back, body, and headaches continued with the fever and chills. Because he was not getting better, he came to the ER on day 6 of illness for further evaluation.

The review of systems was significant for headache, mid back pain, chills and fever, and sweats. At the time of admission, his nausea and vomiting were resolving and he had no diarrhea. There was no swelling of any glands noted by the patient.

Past medical history was significant for a trampoline accident at age 13, requiring several weeks of treatment, including tracheostomy.

He had been vaccinated in July 1998 (prior to travel overseas) against hepatitis A and B, polio, meningococcus, typhoid, and cholera.

The patient was single. His girlfriend did not travel overseas with him. He worked in the lawn mowing business, and spent a lot of time outdoors. He recently quit smoking but drank beer occasionally, and had occasional use of marijuana in the past. He was on no regular medications, except ibuprofen several times a day to deal with the body aches and fever in the 7 days prior to admission. He had no known drug allergies.

He had spent about 6–8 weeks on the west-central coast of Africa about 8–9 months earlier, working for an oil company. Countries he visited at that time included Equatorial Guinea and Gabon.

While in Africa, he had spent time on and off shore, including outdoors, also during dusk hours. He was not ill during his stay in Africa. His round trip took him through Europe. Mefloquine had been prescribed for the patient to be taken once weekly as malaria prophylaxis. However, he took it only during his stay in Africa, but failed to complete the prophylaxis on return to the USA.

A hematology technologist reviewing the peripheral blood smear (Wright's stain for differential count) noted intracellular inclusions, and suspected the presence of parasites (Fig. 5.2a).

• *What inclusions do you see within the red blood cells?*

Infectious diseases consultation was sought on 5/15/99, the day of admission, to assess the patient for the possibility of malaria. The blood smear was reviewed by the ID physician who agreed with the hematologist's interpretation that malaria parasites were present. No gametocytes were seen on the smear, but intracellular trophozoites (ring forms) were noted. A presumptive diagnosis was made.

On examination, the patient was alert, oriented, and in no acute distress. The vital signs showed the following: BP 102/48, RR 18, HR 76, temperature 99.6 °F

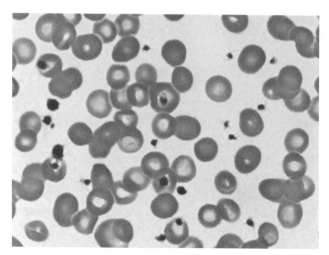

Figure 5.2a Peripheral blood smear (Wright's stain) done on 5/15/99 showing intracellular trophozoites.

Table 5.2a Serial complete blood count (CBC) parameters recorded over several days

CBC/diff	5/15/99	5/16/99	5/17/99	5/21/99
WBC	3.8	4.3	4.5	5.2
H/H	12.1/34.9	11.2/32.7	10.7/31.4	10.3/29.9
MCV	84	86.1	85.0	84
Platelet	32	37	62	276
% PMN	63	52.2	36	54.4
% Band	24	0.00	19	0.00
% Lymph	9	38.0	30	35.0
% Mono	3	9.4	14	9.6
% Eos	0	0.2	0.00	0.4

Eos, eosinophil; H/H, hemoglobin and hematocrit; Lymph, lymphocyte; MCV, mean corpuscular volume; Mono, monocyte; PMN, polymorphonuclear leukocyte; WBC, white blood cell.

(T_{max} in the ER was 103.5 °F); height 6′2″, and weight 190 pounds. The head and neck exam showed no jaundice, no oral lesions, and no adenopathy. The heart, lung, abdomen, upper and lower extremities were unremarkable. Neurologic exam was normal while the skin showed a good suntan, with no rash noted. He had no palpable lymphadenopathy.

Laboratory and other investigations

Table 5.2a shows CBCs with differentials recorded over several days. Note that the WBC count was not elevated, and the platelet count on admission was very low at 32,000/µL. There was no eosinophilia noted on peripheral smear differential.

On 5/15/99, the following laboratory findings were noted: chemistries showed a sodium level of 133, potassium 3.1, glucose 118, BUN/creatinine 13 and 1.1, respectively. The liver tests showed SGOT 25, SGPT 40, alkaline phosphatase 81, total bilirubin 1.9, albumin 3.2, and total protein 6.2 g/dL. The total creatine phosphokinase (CPK) was normal at 43, LDH 800 (normal 313–618 IU/L), and phosphorus was low, 0.8 (normal 2.5–4.9 mg/dL). Blood cultures ×2 were negative. On 5/17/99, the glucose-6-phosphate dehydrogenase (G6PD) level was normal.

Case 5.3   A 57-year-old male with acute-onset fever and chills

A 57-year-old Caucasian male was seen in the emergency department (ED) late on 6/9/15 with fever, chills, fatigue, and body aches. His symptoms had started on that day, less than 12 hours before. His temperature at home was as high as 104 °F. He was brought to the ED by his wife for further evaluation. An airline pilot, he had flown in from the West Coast of the United States back home to the Gulf Coast of Mississippi that same day. Because of his history of international travel, the initial suspicion was for some exotic viral infection. The physical examination was in general unremarkable. He had no jaundice and no rash.

His initial labs showed a WBC of 2.8, H/H 14.3 and 43.7, respectively. MCV was normal at 87.5, but platelet count was low at 86,000. The comprehensive metabolic profile, including liver tests and renal function tests, was unremarkable.

The past history was significant for the following: known aortic stenosis with systolic murmur, allergic rhinitis, and various surgical procedures that included colonoscopy ×3, left Achilles tendon repair, left upper jaw dental implant, tonsillectomy and appendectomy in childhood, as well as right orchiectomy for seminoma in 1978, treated and cured following radiation therapy. He had allergy to penicillin: rash, noted in childhood years.

The family and epidemiologic history were significant for the following: a pilot with an international airline, he was married and had two grown daughters in their late 20s. He did not smoke or drink. No member of the family was ill. For the last 2–3 years he had been flying from a US city to Lagos, Nigeria, typically weekly, 3–4 times a month. These weekly flights are non-stop and he would stay overnight in Nigeria, at an international hotel with air conditioning, and would leave the next day on the trip back to the US. He left on 5/27/15 from Lagos on his last trip back home, 12 days before the ED visit.

- *What differential diagnoses would you contemplate in this patient?*

He was required to take malaria prophylaxis, and had been prescribed daily atovaquone-proguanil. However, he admitted taking the last dose of this drug 1 year earlier, in June 2014.

Following initial evaluation, blood cultures ×2 were done, along with other labs, including malaria smears, thick and thin smears by Giemsa stain.

A presumptive diagnosis of malaria was made and the patient was prescribed an antimalarial agent. Fortunately, he already had some atovaquone-proguanil at home, left over from his previous prescription for prophylaxis. Although he was clinically stable, it was important to start antimalarial therapy as soon as possible. Typically, antimalarial drugs are not usually available on demand at the local pharmacies. However, he was able to start his antimalarial therapy within a few hours of leaving the ED. He was prescribed 3 days of four tablets of atovaquone-proguanil daily, which he completed. He started feeling much better within 24–48 hours of starting the medication.

I spoke with the ED physician and the patient's primary care physician later, and with the patient, to confirm that he was doing well. I reviewed the Giemsa malaria smears (thin smears) shown in Figs 5.3a and 5.3b. The smears show the ring forms (early trophozoites) in normal-sized RB's, as well as two ring form trophozoites in a single RBC.

We requested follow-up malaria smears, to document clearance of the parasitemia.

One week later, the patient was off on vacation with his family, already back to his baseline healthy status.

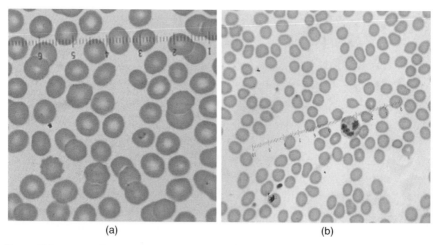

(a)   (b)

Figure 5.3a A. Peripheral blood malaria stain (Giemsa, thin smear) done on 6/9/15 showing intracellular ring forms in normal-sized red cells. B. Segmented PMN noted for size comparison. (*See insert for color information*)

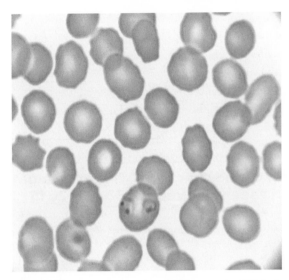

Figure 5.3b Peripheral blood malaria stain (Giemsa, thin smear), 6/9/15, showing intracellular ring forms in normal-sized red cells. One RBC had two ring forms, suggesting *P. falciparum*.

Follow-up visit and post-treatment course

The peripheral smears were reviewed later, confirming the result as compatible with malaria, likely *falciparum* malaria, especially with two ring forms in a single normal-sized RBC.

The follow-up Giemsa smear done 6 days later on 6/15/15 (before his vacation trip) showed no malaria parasites. The platelet count had recovered back to normal at 202,000. The WBC was 4.4 (up from 2.8). The comprehensive metabolic profile was still normal, including the liver and renal function tests.

He was seen 3 weeks later in the office on 7/2/15 for a scheduled follow-up visit. He was completely asymptomatic. He admitted that he had not taken the prescribed prophylaxis for malaria. There had been no unusual exposures that he was aware of, including to mosquitoes, but he did spend some brief time in Nigeria outside his air-conditioned hotel room.

His physical exam was generally unremarkable, except for a systolic ejection murmur, grade 3/6, consistent with the known aortic stenosis. He was well-built and fit. He had no jaundice and no rash. His neurologic exam was normal.

Results of the *Plasmodium* species PCR test (done at a reference laboratory) were received just before the scheduled office visit. The test confirmed that *P. falciparum* was positive, while the test was negative for the other species: *P. knowlesi*, *P. malariae*, *P. ovale*, and *P. vivax*. This was compatible with the clinical impression and the Giemsa thin smear findings previously described.

We did not order a G6PD test on the patient and he did not receive any additional therapy, such as primaquine, because for *P. falciparum*, the atovaquone-proguanil therapy was adequate, with no need for a "radical cure" of the liver phase infection.

• *Final diagnosis: Uncomplicated P. falciparum malaria.*

The patient had appropriate antimalarial therapy, and responded as expected. He did not require admission to the hospital.

Comparison of the characteristics of five patients with malaria seen between 1998 and 2015

Table 5.3a is a comparison of the characteristics of five patients with malaria seen between 1998 and 2015.

Case discussion

All five patients presented here had high fever (maximum temperature above 103 °F) with chills, in association with head and body aches. They were all relatively young people who had undertaken recent travel, except for case #2, who had left the region of infection acquisition 8–9 months earlier. He did not complete his malaria prophylaxis appropriately. In fact, two of these five patients did not take any antimalarial medication at all (patients #1 and #4), while patients #2 and #3 took the prescribed prophylactic antimalarial medication inappropriately or incompletely. Case #5 was an airline pilot who did frequent trips to Nigeria, in West Africa. He had not taken any antimalarial prophylaxis since June 2014, for 1 year before the onset of his febrile illness with malaria.

Plasmodium falciparum was found in the two patients who acquired their infection from Central America or the Caribbean, while two of those acquiring their infection from West-Central Africa likely had non-*falciparum* malaria. All of these patients recovered without sequelae, and responded appropriately to antimalarial medication. Thrombocytopenia resolved rapidly in all five patients within a few days to 1 week of starting appropriate antimalarial medication.

All five cases of malaria were considered imported into the United States, while two of three cases acquired from Africa were both imported and possibly relapsing or recrudescent. Case #5 was uncomplicated acute acquired *P. falciparum* malaria, imported from West Africa. This patient had not taken the prescribed atovaquone-proguanil in 1 year.

The patient seen in 2006 (case #4) acquired her malaria while vacationing in the Dominican Republic, and became ill after return to the USA. Her illness was complicated by iatrogenic septic thrombophlebitis (methicillin-resistant *Staph. aureus* positive blood culture 4 days after admission) related to contaminated

Table 5.3a Comparison of five cases of malaria seen between 1998 and 2015

Characteristic	Case 1 (Case 5.1)	Case 2 (Case 5.2)	Case 3	Case 4	Case 5 (Case 5.3)
Dates of hospitalization (days)	10/28–11/2/98 (5 days)	5/15–5/17/99 (2 days)	5/3–5/8/01 (5 days)	4/23–4/29/06 (6 days)	6/9/15 (0 day): seen in ED only
Age (years)/sex	15/male	23/male	43/male	39/female	57/male
Major symptoms	Fever, chills, rigors, dry cough, headache	Nausea, vomiting, fever, chills, sweats, head and body ache	Splitting headache, fever, chills, rigors, body aches, dry cough	Fever, chills, head and body aches; some nausea, fatigue, and malaise	Fever, chills, fatigue, body aches
Duration of symptoms before admission	6 days	6 days	2–3 weeks?	2–3 days	One day
Max. temperature recorded (°F)	104.4	103.5	104.5	103.2	103.7
Lowest platelet count noted	31,000	32,000	96,000	22,000	86,000
Region where infection was acquired	Central America	West-Central Africa	West-Central Africa	Caribbean (Dominican Republic)	Nigeria, West Africa
Time from presumptive exposure to illness onset (estimate)	13–19 days (2–3 weeks)	Probably 8–9 months	2–4 weeks	18 days (2–3 weeks)	2–4 weeks

(continued)

Table 5.3a (*Continued*)

Characteristic	Case 1 (Case 5.1)	Case 2 (Case 5.2)	Case 3	Case 4	Case 5 (Case 5.3)
Plasmodium sp. incriminated	P. falciparum	Non-falciparum	Likely non-falciparum	P. falciparum	P. falciparum
Chemoprophylaxis taken	None	Inconsistent or incomplete	Inconsistent or sporadic	None	None
Mode of acquisition	Imported	Imported or relapsing	Imported/relapsing/or recrudescent	Imported	Imported
Outcome/treatment	Survived/chloroquine	Survived/quinine and doxycycline	Survived/quinine and doxycycline	Survived/chloroquine	Survived/atovaquone-proguanil
Other comments	Patient responded rapidly to Rx. Chloroquine therapy was adequate because P. falciparum was confirmed, and there was no chloroquine resistance in the region	Because the Plasmodium species was uncertain, he received additional therapy with primaquine	A merchant marine; made several trips to malarious areas over years. Also received additional therapy with primaquine	Her illness was complicated with thrombophlebitis 4 days after admission (+ BC for MRSA), requiring additional Rx	Malaria was uncomplicated. He cleared the parasitemia rapidly. There was no need for G6PD test or use of primaquine

BC, blood culture; ED, emergency department; G6PD, glucose-6-phosphate dehydrogenase; MRSA, methicillin-resistant *Staph. aureus.*

Figure 5.4a Nineteen cases of malaria, including four among travelers, were reported as acquired on the island of Great Exuma in the Bahamas in May–June 2006. Source: www.cdc .gov/mmwr/preview/mmwrhtml/mm5537a1.htm, accessed February 26, 2016.

percutaneous venous access. This complication prolonged her period of fever, even after appropriate chloroquine therapy. She made a complete recovery.

In 2006, malaria (*P. falciparum)* was confirmed in travelers to Great Exuma, Bahamas, and Kingston, Jamaica, areas where malaria transmission typically did not occur [1]. Our 2006 patient (case #4) was in the region (Dominican Republic) vacationing during that same time period (Fig. 5.4a). *P. falciparum* causes 99% of malaria in Hispaniola (Haiti and Dominican Republic), and has remained sensitive to chloroquine [1].

Only two of three patients who acquired their malaria in Africa were given primaquine, an agent that is active against *P. vivax* and *P. ovale*, species that have exoerythrocytic or liver stages of the *Plasmodium* life cycle [2]. This 14-day treatment, typically given after eradication of the blood parasite, is designed to clear the liver phases of the parasite, and is considered a "radical cure" treatment. This treatment is designed to prevent a relapse of malaria months later, as may have occurred in our patient #2, who became ill with malaria 8–9 months after return to the United States. The patient who had *P. falciparum* malaria (case #5) did not need or require primaquine treatment after treatment of the acute malaria (Fig. 5.4b).

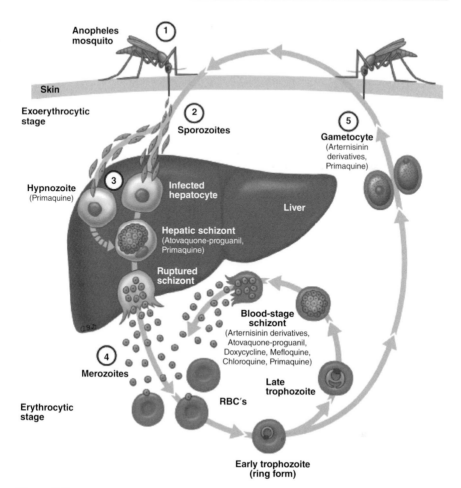

Figure 5.4b

Life cycle of *Plasmodium*. (1) *Plasmodium*-infected *Anopheles* mosquito bites a human and transmits sporozoites into the bloodstream. (2) Sporozoites migrate through the blood to the liver where they invade hepatocytes and divide to form multinucleated schizonts (pre-erythrocytic stage). Atovaquone-proguanil and primaquine have activity against hepatic-stage schizonts. (3) Hypnozoites are a quiescent stage in the liver that exist only in the setting of *P. vivax* and *P. ovale* infection. This liver stage does not cause clinical symptoms, but with reactivation and release into the circulation, late-onset or relapsed disease can occur up to many months after initial infection. Primaquine is active against the quiescent hypnozoites of *P. vivax* and *P. ovale*. (4) The schizonts rupture and release merozoites into the circulation where they invade red blood cells. Within red cells, merozoites mature from ring forms to trophozoites to multinucleated schizonts (erythrocytic stage). Blood-stage schizonticides such as artemisinins, atovaquone-proguanil, doxycycline, mefloquine, and chloroquine interrupt schizogony within red cells. (5) Some merozoites differentiate into male or female gametocytes. These cells are ingested by the *Anopheles* mosquito and mature in the midgut, where sporozoites develop and migrate to the salivary glands of the mosquito. The mosquito completes the cycle of transmission by biting another host.

*There is strong evidence that drugs listed in parentheses are active against designated stage of parasitic life cycle. Primaquine is a blood-stage schizonticide with activity against schizonts of *P. vivax* but not those of *P. falciparum*.

•Reproduced with permission of UpToDate, Inc. from Hopkins H. Diagnosis of malaria. Available at: www.uptodate.com/contents/diagnosis-of-malaria?source=search_result&search=Malaria&selectedTitle=4%7E150, accessed February 26, 2016. (*See insert for color information*)

Comments

Most cases of malaria in the United States are imported, with the largest percentage of cases acquired in Africa, followed by Asia, Central America, and the Caribbean [3,4]. The largest category of US civilians by purpose of travel at the time of malaria acquisition is those visiting friends or relatives [3,4]. The most common species of *Plasmodium* seen in imported malaria cases is *P. falciparum*, 94% of which occur within 30 days of return to the US [4]. Malaria occurring more than 2 months after return to the US is more likely to be due to a species other than *P. falciparum*.

In the 2005 CDC surveillance report [4], approximately 80% of imported malaria cases among US civilians occurred among those who either were not taking prophylaxis or were taking non-recommended prophylaxis for the region to which they were traveling.

Our three patients who acquired their infection in Africa did not take the recommended prophylaxis appropriately, while the two who acquired their infection in Central America or the Caribbean did not even know they were at risk for malaria. All of our patients had significant thrombocytopenia at presentation, with resolution occurring very rapidly after appropriate antimalarial therapy. In three of our five patients, severe thrombocytopenia was a sensitive indicator finding for malarial infection in those non-immune individuals. Two of the patients (cases #1 and #4) were seen in consultation by the hematologist because of the severe thrombocytopenia (before infectious disease consultation was placed). One of them (case #1) actually had a bone marrow biopsy done.

Malaria should be considered in the differential diagnosis of a febrile patient with severe thrombocytopenia in the returning traveler.

Of the seven persons with fatal outcomes noted in the United States in 2005 [4], none had taken prophylaxis, and substantial delays occurred in their seeking care or in diagnosis and treatment, or both.

Thus, it is very important to take appropriate prophylaxis when traveling to a malarious area, to seek help early, and to consider malaria in the differential diagnosis of fever in the returning traveler.

Lessons learned from these cases

- Malaria should be considered in the differential diagnosis of fever in the returning traveler.
- Appropriate and adequate prophylaxis is necessary to prevent malaria.
- Severe thrombocytopenia may be an indicator finding for malarial infection in the non-immune individual with fever, in the appropriate setting.
- Malaria is a serious disease that responds well to early and appropriate therapy.
- Appropriate travel history is important in making the diagnosis of malaria.
- Knowledge about the area where the disease was acquired is important in determining initial appropriate therapy.

References

1 Malaria – Great Exuma, Bahamas, May–June, 2006. Morbidity and Mortality Weekly Report 2006; 55(37): 1013–16. Available at: www.cdc.gov/mmwr/preview/mmwrhtml/mm5537a1 .htm, accessed February 26, 2016.

2 Hopkins H. Diagnosis of malaria. Available at: www.uptodate.com/contents/diagnosis-of-malaria?source=search_result&search=Malaria&selectedTitle=4%7E150, accessed February 26, 2016.

3 Malaria surveillance – United States, 2010. Morbidity and Mortality Weekly Report 2012; 61(2): 1–22. Available at: www.cdc.gov/mmwr/pdf/ss/ss6102.pdf, accessed February 26, 2016.

4 Malaria surveillance – United States, 2005. Morbidity and Mortality Weekly Report 2007; 56(SS06): 23–38. Available at: www.cdc.gov/mmwr/preview/mmwrhtml/ss5606a2.htm, accessed February 26, 2016.

Further reading

Breman JG. Clinical manifestations of malaria. Available at: www.uptodate.com/contents/ clinical-manifestations-of-malaria?view=print, accessed February 26, 2016.

Laboratory identification of parasites of public health concern. Available at: www.dpd.cdc.gov/ dpdx/HTML/Para_Health.htm, accessed February 26, 2016.

Olliaro P, Cattani J, Wirth D. Malaria, the submerged disease. JAMA 1996; 275(3): 230–3.

CHAPTER 6

Gulf Coast Tick Rash Illness*

Case 6.1 Woman with rash and prolonged eschar following a tick bite

A 63-year-old Caucasian female was admitted in April 2008 because of rash and body aches (Table 6.1a, case #2). Six days prior to admission (PTA), she had noted the onset of severe head and body aches, followed by exacerbation of arthralgias. Four days PTA, she noted the onset of a rash, beginning in the upper torso of the chest and spreading to involve other body parts within 24 hours. Two days later (and 2 days PTA), she came to the emergency room (ER) and was given doxycycline, on suspicion for Rocky Mountain spotted fever (RMSF). She returned to the ER 2 days later because she was not feeling any better, and was thus admitted.

Infectious diseases consult was sought the next day for the body aches, headaches, and generalized rash.

Epidemiology and review of systems discovered that this lady spent a lot of time outdoors, gardening and handling flowers. In late March, 2008 (10 days prior to the onset of symptoms), she had noted something "like a small blood clot" on her right ankle. Her husband picked it off, and saw what appeared to be "legs" or "feet" under the microscope. He crushed this unusual tick with a piece of stick.

Headaches and other symptoms started 10 days later, with rash noted 12 days after this "unusual tick" was removed.

• *What are the common tick-borne diseases to consider in this patient at this time?*

She lived with her husband in a small rural town in southern Mississippi. They did not have a dog and nobody else in the household was ill.

The past medical history was significant for type 2 diabetes mellitus with neuropathy, and degenerative joint disease involving the hands, knees, and back. She also had hypertension and hypercholesterolemia. She had smoked 1–2 packs of cigarettes a day for 40 years.

She was allergic to codeine (nausea and vomiting).

Medications PTA were as follows: doxycycline 100 mg BID started 2 days PTA; others included rosiglitazone, alendronate plus D, meloxicam, ezetimibe/simvastatin, gabapentin, propranolol LA, and lisinopril hydrochlorothiazide.

Cases in Clinical Infectious Disease Practice: Obtaining a Good History from the Patient Remains the Cornerstone of an Accurate Clinical Diagnosis: Lessons Learned in Many Years of Clinical Practice, First Edition. Okechukwu Ekenna.
© 2016 John Wiley & Sons, Inc. Published 2016 by John Wiley & Sons, Inc.

Table 6.1a Characterization of patients with suspected *Rickettsia parkeri* infection.

Characteristic (date seen)	Case 1 (4/07)	Case 2 (case 6.1) (4/08)	Case 3 (7/08)	Case 4 (9/08)	Case 5 (case 6.2) (8/12)
Age/sex	72 year C/M	63 year C/F	45 year C/F	65 year C/M	58 year C/M
Site of tick bite	R. groin	R ant. ankle	Abdomen and back	L. chest and LE	R. groin
Duration of eschar	>15 days	>17 days	>15 days	Not documented	≥21 days
Tick bite to onset of symptoms	7 days	10 days	3 days	14 days	14 days
Duration of symptoms	>8 days	>7 days	>14 days	>10 days	≥10 days
Bite to onset of rash	13 days	12 days	14 days	>14 days	≥14 days
Duration of rash	Unknown	>7 days	>5 days	>5 days	≥5 days
Extent of rash: type	LE, UE, torso, palms/soles: erythematous-macular	Generalized (no palms or soles): macular > nodular	Ant. chest, back, LE, esp. thighs: few scattered malar	Generalized, more intense in LE (no palms or soles): macular petechiae	Generalized: torso, UE, and LE (no palms or soles); papular > macular
Symptoms experienced by patients	Fever, chills, H/A, rash; T_{max} 102.4	Gen. body and H/A, arthralgia, chills, nausea; T_{max} 99.6	Severe H/A, body, neck and joint pains, low-grade temp, chills, fatigue	Body aches, bitemporal H/A, fever and chills, rash; T_{max} 103.5	Joint and body aches, H/A, fever, chills, fatigue, dry cough; T_{max} 102.5
Laboratory diagnosis*	*R. parkeri* detected in attached *Amblyomma maculatum* tick by PCR	RPA-IgG IFA = 1:1024 on 5/7/08	RPA-IgG IFA: 7/9/08: <32 8/18/08: 256 CSF negative	RPA-IgG IFA: 9/25/08: 1:1024 10/17/08: 1:64 CSF negative	Eschar bx: IHC positive for SFGR and PCR positive for *R. parkeri* CSF negative

Underlying diseases	NHL, recent chemo 3 d PTA, neutropenia, DM	DM, DJD, overweight, heavy tobacco use	PM seizure, kidney stones, overweight	Gout, DJD, arthritis, overweight	HCV 20 years ago, appendectomy, overweight
Treatment received	Vancomycin, cefepime, doxycycline	Doxycycline, ceftriaxone	Doxycycline	Doxycycline, ceftriaxone	Doxycycline
Outcome	Full recovery	Full recovery	Full recovery	Full recovery	Full recovery
Exposure history and epidemiology	Rider lawn mowing; tick in R. groin	Gardening, handling flowers; tick in R. ankle	Realtor visiting rural site; 6–8 ticks on back and abdomen	Weeding and planting grass in deer camp; four ticks L. chest and LE	Tick in R. groin. Rider lawn mowing. Walking in tall grass in old property

*A confirmed case of *R. parkeri* rickettsiosis is defined by PCR amplification of *R. parkeri* DNA from a clinical specimen. A probable case is defined as an epidemiologically and clinically compatible illness with a supportive serologic or immunohistochemical test result, through use of group-specific assays for SFGR, or PCR amplification of *R. parkeri* DNA from an attached tick [3].

DJD, degenerative joint disease; DM, diabetes mellitus; H/A, headache; IFA, indirect immunofluorescent antibody assay; IHC, immunohistochemistry; LE, lower extremity; NHL, non-Hodgkin's lymphoma; PCR, polymerase chain reaction; PTA, prior to admission; RPA, *Rickettsia parkeri*; RR, *Rickettsia rickettsii*; SFGR, spotted fever group rickettsiae; UE, upper extremity.

On examination, she was alert and oriented. Her vital signs were as follows: blood pressure (BP) 106/56, respiratory rate (RR) 18, heart rate (HR) 75, temperature 99.6°F; height 5'5", weight 182 pounds. The head and neck exam showed nothing significant other than prosthetic dentures. The heart, lung, and abdominal exams were unremarkable. The extremities, gross neurologic exam, and lymphatics were negative. The skin showed diffuse, macular and papular rash on the trunk (front and back), bilateral upper and lower extremities, but not on the soles or palms. There was a bite eschar on the right anterior ankle still present 17 days after the tick bite (Fig. 6.1a).

Laboratory findings showed the following: white blood cell (WBC) count 6800/μL, hemoglobin and hematocrit (H/H) 13.4/39.4, respectively, mean corpuscular volume (MCV) 96, platelet count 320,000. Differential count: 65.8% granulocytes, 20.5% lymphocytes, and 12.8% monocytes. Chemistries showed a sodium value of 142, potassium 3.9, glucose 105 mg/dL, and BUN/creatinine 8/0.8, respectively. The liver transaminases were normal, albumin was low at 3.0, and cholesterol and triglyceride values were within normal range. Erythrocyte sedimentation rate was 50 mm/hour. Blood cultures were negative.

- *Which tick-borne diseases are associated with a rash and persistent or prolonged tick bite eschar?*

The ID consultant requested that additional serology be sent to the CDC for further testing. This request was based on knowledge that there were other spotted fever rash illnesses, especially in the south-eastern United States and on the Gulf Coast of Mississippi.

The patient was discharged home 2 days after admission (1 day after ID consult) on continued oral doxycycline (ceftriaxone was discontinued), to be followed up as an outpatient.

She was seen in the office 3 weeks later, at which time convalescent serology was obtained and sent to the CDC. She had completed about 2 weeks of doxycycline therapy. The skin rash was faded but not completely gone. By the next visit 1 month later (2 months after discharge from the hospital), the rash was completely resolved and she was back to her baseline.

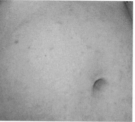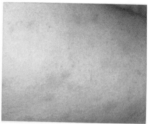

Figure 6.1a Tick bite eschar in anterior right ankle, and macular papular rash on abdominal torso and upper back (reproduced with permission of the Mississippi State Medical Association). (*See insert for color information*)

The diagnosis was confirmed with the convalescent serology from the CDC a few days before that office visit. The convalescent titer for *Rickettsia parkeri* was 1:1024, with the corresponding titer for *R. rickettsii* (RMSF) much lower (1:256). Details are noted in Table 6.1a.

Case 6.2   A 58-year-old male with generalized body aches and fever

A 58-year-old Caucasian male was admitted through the ER in August, 2012 with a 7-day history of malaise, joint aches, cold sensation, and fatigue (Table 6.1a, case #5). He also had a non-productive cough, chest and body aches, and fever.

Review of systems was significant for chest, joint, and body aches, including the elbows, knuckles, and hands, with symptoms ongoing for 7 days PTA. Headache was prominent on admission but was improved 3 days later. The maximal temperature was 102.5°F.

Physical examination was remarkable for a temperature of 100.1°F; multiple red non-pruritic papules were generally distributed over the torso, abdomen, back, upper and lower extremities, including the dorsi of the feet, and there was an estimated 1.3 cm eschar in the right inguinal area. The remainder of the examination was otherwise unremarkable. Photographs of the skin showing the right groin eschar and papular nodular rash noted in many body parts are shown in Fig. 6.2a.

Computed tomography (CT) scans of the head and the chest x-ray were normal. Myocardial perfusion single photon emission computed tomography (SPECT) was normal with a left ventricular ejection fraction of 75%. Abdominal ultrasound showed fatty infiltration of the liver. Fluoroscopic lumbar puncture revealed reddish tinged cerebrospinal fluid (CSF) and was considered traumatic; CSF glucose was 68, protein 52, red blood cells (RBC) 2106, and WBC 3/mm³; differential count: 16% polymorphonuclear leukocytes (PMNs),

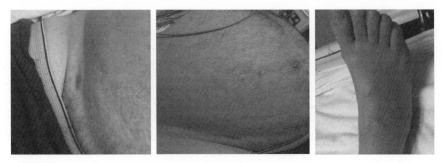

Figure 6.2a Tick bite eschar right groin and papular nodular rash on abdomen and foot dorsum (reproduced with permission of the Mississippi State Medical Association).

⁄o lymphocytes, and 48% mesothelial cells. The CSF gram stain showed no organisms; the culture was negative. Two sets of blood cultures were negative. The patient was started on ceftriaxone and doxycycline intravenously.

With negative blood and CSF cultures, infectious disease consult was sought 4 days after admission to address the fever of unknown origin.

The patient recalled a tick bite 3 weeks PTA while walking in a field of tall grass. His medical history included a diagnosis of hepatitis C for which he had been treated, although he was not sure whether there was partial or complete remission. Complete blood count and chemistries on admission were unremarkable. Liver tests showed aspartate aminotransferase (AST) (serum glutamic oxaloacetic transaminase [SGOT]) of 73 U/L, alanine aminotransferase (ALT) (serum glutamic pyruvic transaminase [SGPT]) 91 U/L with normal alkaline phosphatase, albumin, total protein, and bilirubin levels. Four days later, the AST had dropped to 48 U/L and the ALT to 57 U/L.

• *What would be your next diagnostic step to arrive at a diagnosis in this patient?*

An excisional biopsy of the right groin eschar was obtained and evaluated at CDC by histopathology, immunohistochemistry, and polymerase chain reaction (PCR) tests. The histopathology and immunohistochemical stains of the biopsy are shown in Fig. 6.2b.

The patient was discharged home after the biopsy was done, on oral doxy-cycline. At follow-up visit 2 weeks later, the patient had fully recovered and his rash had resolved.

Case discussion

The two patients described above presented with tick-borne infections that caused symptoms similar to RMSF, although milder. During the period that these patients were seen, there was no commercial serologic test available to identify the causative agent of this rash illness. We sent the patients' serum (cases #1–5 in Table 6.1a) and eschar biopsy (case #5 in Table 6.1a) to the CDC for confirmation.

The clinical characteristics of the five cases seen between 2007 and 2012 are summarized in Table 6.1a. The second case presented here (case #5) was defini-tively confirmed, while the other four cases were very probably cases of Gulf Coast tick rash illness. Three of the five patients (cases #3, #4, and #5) had lumbar puncture as part of the work-up of the febrile rash illness.

Parker was the first to describe the infectious agent in the vector tick, *Ambly-omma maculatum*, in 1939 [1]. This relatively new pathogenic *Rickettsia, Rickettsia parkeri*, is named after him. Infection with *R. parkeri*, a newly determined human pathogen [2], has been reported throughout the south-eastern US, with a large portion of confirmed or probable cases coming from Mississippi [3]. A more recent paper by Paddock et al. has summarized in more clinical and pathologic

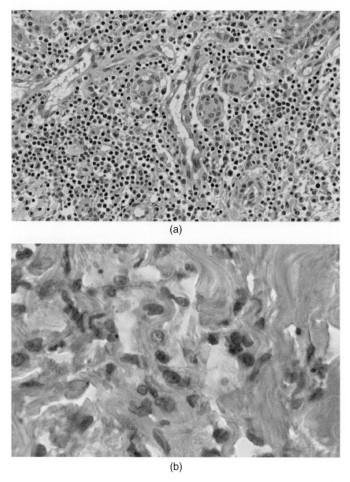

(a)

(b)

Figure 6.2b Histopathology and immunohistochemical staining for spotted fever group rickettsiae (SFGR) of right groin eschar of case #5. A. Hematoxylin and eosin stain showing diffuse lymphocytic perivascular inflammation: original magnification ×100. B. Immunohistochemical stain for SFGR (red). Original magnification ×258 (reproduced with permission of the Mississippi State Medical Association). (*See insert for color information*)

detail some of the major features of this new addition to the spotted fever group (SFG) of rickettsiae [3].

The Gulf Coast tick rash illness has many other synonyms that include:

- Gulf Coast tick rash illness (after the Gulf Coast tick vector, *Amblyomma maculatum*)
- American boutonneuse fever (Goddard) [4]
- spotted fever rickettsiosis
- spotted fever transmitted by the Gulf Coast tick
- *Rickettsia parkeri* rickettsiosis
- *Amblyomma maculatum* infection.

We suspect that there may be many more of these patients than are reported because there is currently no readily available commercial serologic test to detect *R. parkeri*, and also because of the low index of suspicion for *R. parkeri* infections by clinicians. Many such cases likely did not generate infectious diseases consultation, or the patients may have been treated presumptively for RMSF.

A review of the clinical characteristics in this small number of patients suggests the following.

- The duration of the tick bite eschar was prolonged (longer than 14 days in our five patients).
- Generalized body and headaches were prominent symptoms, while gastrointestinal symptoms (nausea) were reported in only one of five patients presented (case #2).
- Time from tick bite to onset of symptoms ranged from a few days to 2 weeks.
- Time of tick bite to onset of rash in our five patients was between 12 and 14 days.
- Rash was usually present for 5 days or longer.
- Illness symptoms lasted for more than 7 days in all of our five patients.
- All five patients made a full recovery from the illness (zero mortality).
- The illness was generally milder than that usually reported for RMSF.

A comparison of selected clinical features of three spotted fever group rickettsioses was presented in the paper by Paddock et al. [3]. *R. parkeri* was compared with RMSF and rickettsial pox. Rickettsial pox (caused by *R. akari*) is typically an urban disease transmitted by the mouse mite [5,6], while *R. parkeri* infection is usually acquired via a tick bite, with outdoor or rural exposure. Both of these infections are transmitted through a bite, often leaving an eschar, although rickettsial pox is not caused by a tick but by a rodent (mouse) mite [5,6].

None of the 12 patients with *R. parkeri* reported in the Paddock paper died, and only one (8%) had any gastrointestinal (GI) symptoms (nausea and vomiting) reported. The mortality was higher and the percent hospitalization was also higher in RMSF patients than in those with *R. parkeri*. All of our five patients reported here survived.

Lessons learned from these cases

- The Gulf Coast tick rash illness appears to be clinically similar to but milder than RMSF.
- The clinical spectrum of the disease is still incompletely defined in this new SFG rickettsia.
- The tick bite eschar appears to remain for a prolonged period, and may serve as a useful clue or diagnostic sign. This factor, however, will need to be confirmed in a larger number of patients.

- Outdoors exposure to ticks in grass or lawn was characteristic in all of the affected patients, suggesting a rural rather than an urban disease.
- An accurate and rapid diagnosis would likely lead to the avoidance of invasive procedures like lumbar puncture, as was done in three of the five patients in this series.
- Doxycycline (tetracycline) appears to be an effective therapy for this illness, as it has been for the other SFG rickettsioses.
- It is anticipated that once commercial serologic or other tests become widely available, we would likely see many more cases of *R. parkeri* reported.

Acknowledgment

The cases presented in this section were adapted from a recent original publication by the author, with permission from the *Journal of the Mississippi State Medical Association* [7].

References

1 Parker RR, Kohls GM, Cox GW, Davis GE. Observations on an infectious agent from *Amblyomma maculatum*. Public Health Report 1939; 54: 1482–4.

2 Paddock CD, Sumner JW, Comer JA, et al. *Rickettsia parkeri:* a newly recognized cause of spotted fever rickettsiosis in the United States. Clinical Infectious Diseases 2004; 38: 805–11.

3 Paddock CD, Finley RW, Wright CS, et al. *Rickettsia parkeri* rickettsiosis and its clinical distinction from Rocky Mountain spotted fever. Clinical Infectious Diseases 2008; 47: 1188–96.

4 Goddard J. American boutonneuse fever: a new spotted fever rickettsiosis. Infections in Medicine 2004; 21: 207–10.

5 New York City Department of Health and Mental Hygiene. Rickettsial pox (*R. akari*): Bureau of Communicable Disease. Available at: www.nyc.gov/html/doh/html/diseases/cdrick1.shtml, accessed February 29, 2016.

6 Paddock CD, Zaki SR, Koss T, et al. Rickettsialpox in New York City: a persistent urban zoonosis. Annals of the New York Academy of Sciences 2003; 990: 36–44. Available at: www.ncbi.nlm.nih.gov/pubmeds/12860597, accessed February 26, 2016.

7 Ekenna O, Paddock CD, Goddard J. Gulf Coast tick rash illness in Mississippi caused by *Rickettsia parkeri*. Journal of the Mississippi State Medical Association 2014; 55(7): 216–19.

CHAPTER 7

Infectious Diseases Associated with Trauma and Outdoor Activities

Introduction

Infectious diseases associated with trauma and outdoor activities are very varied, but typically involve a breach of the integument or mucous membrane. Such injuries may be caused by abrasions, bites, cuts, punctures, burns or caustic injury, damage of tissue through traumatic crush injury, as in accidents or gunshot, iatrogenic through surgery, puncture, needle injury or injection, and subsequent contamination of damaged skin, other body surfaces, or mucosal surfaces with dirt or microbial flora from the site of injury (normal flora), the surrounding area, or from a distant site.

We will first present two cases of infection resulting from local injury to the skin, and then briefly address some of the general characteristics of infectious diseases associated with trauma and the outdoors.

Case 7.1   A 30-year-old man with thumb infection and forearm nodules

A 30-year-old Caucasian male was admitted with a 3-week history of pustular lesion of the right thumb in late September, 2004.

He first noted a tiny pustular lesion on the right thumb, which he scratched, but found only minimal, thick, whitish pus. Over the next week, the lesion got larger. Doxycycline was prescribed by a healthcare provider, but the pustules progressed and scabbed over. Over the next 2 weeks, nodular, red, bumpy lesions were noted in the subcutaneous tissue of the proximal thumb and distal right wrist. Eventually, streakiness was noted, involving the forearm and moving towards the axilla. In the week before admission, the initially reddish nodules were now skin color. In the 3 weeks prior to admission, the patient visited the emergency room (ER) and workplace clinics, and received various antimicrobials including ciprofloxacin, cephalexin, levofloxacin, and trimethoprim/ sulfamethoxazole. He did not improve, and so was admitted to the hospital.

Cases in Clinical Infectious Disease Practice: Obtaining a Good History from the Patient Remains the Cornerstone of an Accurate Clinical Diagnosis: Lessons Learned in Many Years of Clinical Practice, First Edition. Okechukwu Ekenna.
© 2016 John Wiley & Sons, Inc. Published 2016 by John Wiley & Sons, Inc.

The review of systems was significant for no fever or chills. He had no nausea, vomiting, headache, abdominal, or any constitutional symptoms. He had no significant or severe pain.

His medical history included chronic sinusitis and respiratory allergy, with two sinus surgeries at age 12 and 27 years. He had a fish fin puncture 5 years earlier, requiring surgical repair of damaged tendon. Tetanus toxoid was given 5 years earlier.

This man was the plant manager of a shrimp and fish packing company. He handled squid, shrimp, and fish products. The company packaged its marine products in wooden crates. However, he was unaware of any puncture injuries to the hand. He rarely drank alcohol and did not smoke. He was on no medications, except the antibiotics prescribed prior to admission to hospital. There was no family member ill, nor was there an outbreak at work of a similar illness.

Infectious diseases consult was sought on 9/29/04, 1 day after admission, to help with the diagnosis of the pustular thumb lesion that had failed outpatient antimicrobial therapy.

On examination, the patient was alert and oriented, and in no acute distress. The vital signs were unremarkable: blood pressure (BP) 136/75, pulse rate (P) 85, respiratory rate (RR) 20, and temperature 98 °F. His height was 5'10" and weight 180 pounds. The head, neck, heart, lung, abdominal, lymphatic, and neuropsychiatry examinations were unremarkable. The extremities were unremarkable, except the right upper extremity. There was a scabbed and peeling, dry, cutaneous pustule in the distal radial right thumb area, with surrounding cellulitis. Also noted were 3–4 small, palpable, red, raised nodular lesions that were only minimally tender to palpation, in the distal radial wrist, with swelling involving the ball of the right hand. Photographs of the wrist and thumb are shown in Fig. 7.1a. Several palpable nodules were noted on the volar forearm, up to just below and distal to the elbow. The left arm was unremarkable.

- *What are the likely differential diagnoses?*
- *What test or next step would you recommend to help in arriving at a diagnosis?*

The laboratory findings showed a normal complete blood count (CBC), with a white blood cell (WBC) count of 5.8 and normal differentials. Sedimentation rate was 26 mm/hour (normal 0–20). Routine culture of the scabbed wound showed a positive culture for coagulase-negative *Staphylococcus*.

The infectious diseases consultant recommended a biopsy of one of the forearm nodules. Tissue biopsy was obtained by the surgeon on that day, of both the right thumb scabbed lesion and a forearm subcutaneous nodule.

The histopathology of the right thumb subcutaneous tissue and the subcutaneous nodule from the right inner forearm both showed acute and chronic inflammation. There were foreign body type giant cells present. Acid-fast bacilli (AFB) and Grocott's methenamine silver (GMS) special stains were negative for organisms. The routine bacterial cultures were reported positive for *Staphylococcus epidermidis*.

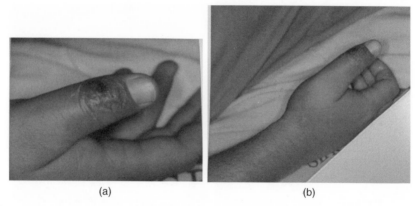

(a)   (b)

Figure 7.1a A. Right thumb: scabbed pustular area with skin peeling. B. Inflammation involving both the thumb and radial dorsal ball of the hand. (*See insert for color information*)

Case discussion

This patient had a pustular lesion on the right thumb that did not respond to 3 weeks of outpatient antibiotic therapy, and hence he was admitted to the hospital. In addition, he had multiple subcutaneous nodular lesions in the forearm, associated with the scabbed abscess of the distal thumb, and inflammation noted in the distal wrist and proximal thumb. This was an indolent infection or process, since he had no fever or severe systemic illness or pain.

The initial culture of coagulase-negative *Staphylococcus* reported 1 day after admission (subsequently confirmed to be *Staphylococcusepidermidis*) was not considered the cause of the infection. A biopsy was recommended for diagnosis. Both the distal thumb initial lesion and a subcutaneous nodule of the forearm showed similar histopathology, with negative AFB and fungal stains. The gram stain showed rare WBCs and no organisms.

The initial growth on culture media from both biopsies was reported on day 6 (10/5/04), and confirmed positive at 2 weeks (10/13/04), with preliminary organism identification. A tentative diagnosis was made, but final confirmation and speciation required additional tests on the positive cultures.

• *Can you guess what was cultured and how did you come to that conclusion?*

The clinical presentation and findings, including the epidemiologic exposures, suggested two likely differential diagnoses: *Mycobacterium marinum* infection and sporotrichosis. This patient was in the seafood business and handled fish, shrimp, and other marine organisms regularly in his line of work. Although he could not recall a recent puncture wound, such an event is often missed by many in the business, since it is not considered unusual unless there is visible injury or bleeding. Sporotrichosis would present a similar clinical picture, often with subcutaneous nodules, as was noted in this patient. Only a biopsy and culture would resolve this issue.

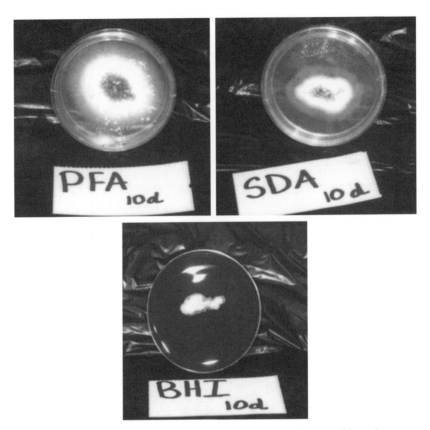

Figure 7.1b Growth of mold on potato flake (PFA), Sabouraud (SDA), and brain heart infusion (BHI) agars on day 10 of subculture of isolated mold from the right arm biopsy: front view of agar plates.

The AFB cultures turned out to be negative, but the fungal cultures were positive on Sabouraud dextrose agar (SDA), with subculture of the mold made to potato flake agar (PFA), as well as brain heart infusion (BHI) agar, in order to identify a "yeast phase," if present. Fig. 7.1b shows 10-day-old growth of subculture of the organism on PFA, SDA, and BHI agar.

- *Can you comment on why the BHI agar was used in this instance?*
- *Final diagnosis: Sporotrichosis of the right thumb, with associated subcutaneous nodules in the right forearm.*
- *Organism isolated: Sporothrix schenckii.*

Comments

The biopsied specimens were placed on SDA and Mycocell agar slants and incubated at 30°C on 9/29/04. Growth on slants was observed on day 6. This growth was subcultured to SDA, BHI/with blood (at 37°C), and PFA agar plates, and incubated at 22 °C (room temperature). Slide cultures were made on day 7 (10/12/04). Identification was made based on colony morphology, rate of growth, and microscopic morphology, including yeast phase conversion on BHI/with blood agar, incubated at 37 °C. *Sporothrix schenckii* is a dimorphic mold.

Key to the microbiologic diagnosis was the growth and cultural characteristic of *Sporothrix schenckii* as a thermally dimorphic fungus (like *Histoplasma*, *Blastomyces*, and *Coccidioides*). BHI agar with blood incubated at 37 °F produced a yeast-like culture, while culture on SDA and PFA at 25–30 °C (room temperature equivalent) showed moldy growth characteristics.

Finally, tease mounts were used to identify the microscopic characteristics of the mold phase of the organism: septate hyphae bearing unicellular conidia on denticles typical for *Sporothrix schenkii*, observed with the use of lactophenol cotton blue stain.

Sporotrichosis infection is usually acquired through exposure to soil, wood (e.g. puncture through splint), or plant material. People involved in landscaping, farming or rose gardening may have a particular professional exposure to this organism [1]. This patient's infection was most likely acquired through skin puncture injury.

Lessons learned from this case

- Sporotrichosis is a subacute or chronic infection that most commonly involves the skin and subcutaneous tissue, and then the lymphatics.
- It is easily confused with *M. marinum* infections, because of clinical similarity and the presence of subcutaneous nodules in the lymphocutaneous variety. This patient handled fish in his job and therefore *M. marinum* was high on the list until the culture returned positive for fungus (sporotrichosis).
- As with *M. marinum*, the skin infection is often indolent and persistent.
- Routine bacterial cultures were negative. That fact should lead to further questions.
- Appropriate fungal cultures must be done otherwise the diagnosis will be missed.
- Skin abscesses that do not respond to usual antibacterial treatment should be further assessed clinically, and with additional fungal and mycobacterial cultures.
- Although the infection was most likely acquired through a skin puncture injury, the patient had no recall of that event and it was not initially evident on patient evaluation.

Reference

1 Kauffman CA. Basic biology and epidemiology of sporotrichosis. Available at: www.uptodate
   .com/online/content/topic.do?topicKey=fung_inf/15174&view=print, accessed February 29,
   2016.

Further reading

Kauffman CA, Bustamante B, Chapman SW, Pappas PG. Clinical practice guidelines for the management of sporotrichosis: 2007 update by the Infectious Diseases Society of America. Clinical Infectious Diseases 2007; 45(10): 1255–65.

Case 7.2   Severe infection and sepsis following recreational fishing

A 60-year-old Caucasian male was admitted through the ER in mid-June, 2014 with fever, chills, and body aches. One day prior to admission, he had gone to a coastal island with his family to fish and grill on the beach. Before setting out on the trip that morning, he had brushed his leg against a canvas tent and lacerated his right mid shin while loading the boat for the trip.

He fished and waded into the water, knee and waist deep, in and out all day. He stayed out on the beach for about 6.5 hours, fishing and grilling with the family. He returned home around 1700 hours, showered, and covered the laceration with Band-Aid, after applying topical bacitracin ointment, and then went out.

He went to bed later, but woke up early the next day around 4.00 am with acute onset of chills and fever, and what he thought was sunburn.

He developed body aches and shakes, and 2 hours later he called the ambulance (emergency medical service: EMS) to get him to the hospital ED because he felt he could not drive. By 7.00 am he said the wound appeared dark and that was when he first noted that there was a problem with his leg. At this time he had generalized malaise, diffuse body aches, chills, and fever.

In the ED, the initial complaint was "sunburn and chills." The temperature in the ED quickly rose from 98.2 °F to 101.7 °F. He was described as sunburned, critically on his head and shoulders, and was noted to have a laceration to the right calf that was slightly erythematous. The blood pressure dropped from 118/47 to 94/44. He received intravenous fluids and ceftriaxone. He was deemed to have sepsis and was quickly worked up and admitted to the medical intensive care unit (MICU). The initial working diagnosis was "sepsis and UTI" (he had hematuria on admission). The source of infection was unclear, but the leg wound was suspected. Fig. 7.2a shows a photo of the right leg laceration wound and bruise taken in the ED on the day of admission.

The past medical history was significant for degenerative joint disease, coronary artery disease, hypertension, and hyperlipidemia. Atrial fibrillation was noted in 2013; deep vein thrombosis in the right leg and bilateral pulmonary embolus were noted in August, 2010, following left ankle surgery. He was on anticoagulation therapy (rivaroxaban and aspirin) until the admission. The rivaroxaban was held because of the hematuria noted in the ED on admission. Surgical procedures included the following: coronary artery bypass graft

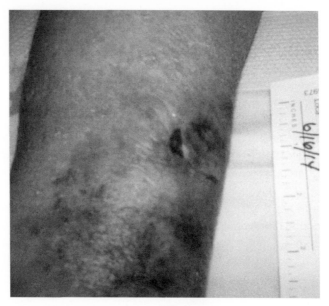

Figure 7.2a Photo taken on the day of admission, 6/16/14, 1 day post laceration of right medial mid-distal shin. Size of blister bruise was noted the next day to be 12.5 × 15 × 0.2 cm (reproduced with permission). (*See insert for color information*)

surgery ×3 in 2013; right inguinal hernia repair ×2 in 1978 and 1990; skin graft right elbow area following heater burn injury at age 14 years; lithotripsy in January, 2014 for right uretero-vesical junction stone, with incomplete stone removal; and appendectomy many years earlier. He was overweight.

Eight days after admission, on 6/24/14, infectious disease consultation was sought to recommend outpatient antimicrobial and other management. The physical examination on that day showed that the patient was alert and oriented, and in no acute distress. The vital signs were as follows: BP 128/69, P 52, RR 18, temperature 97.4 °F; height 6′2″, weight 290 pounds. Head and neck showed seborrheic dermatitis. Heart and lung exam were unremarkable. Abdomen was obese. Lymph nodes were not enlarged. The upper extremities showed a midline catheter in place (right arm), and a peripherally inserted central catheter (PICC) line in the left upper arm. The right leg below the knee was wrapped up in compression dressing (reassessed later). The skin showed light sunburn on the back, seborrheic dermatitis of the face, and a few ecchymotic changes in the forearms.

- *What likely differential diagnoses would you consider for the sepsis?*

Hospital course, laboratory, and other parameters

Blood cultures ×2 on admission on 6/16/14 were reported positive for gram-negative bacilli, with the initial report of positive culture called 11 hours after blood draw. The right leg wound culture obtained on 6/16/14 was also positive for a gram-negative bacillus. The blister and bruise size of the right leg cellulitis was described by the wound care nurse as 12.5 × 15 × 0.2 cm on 6/17/14, 1 day after admission (see Fig. 7.2a).

Several of the patient's laboratory parameters obtained during the hospital admission are noted in Tables 7.2a, 7.2b, and 7.2c. Significant leukocytosis was noted 24 hours after admission (WBC 27.0). Creatine kinase (CK)-MB was normal at 3.2 (normal range 0.5–3.6 ng/mL), as was the troponin I level. The total creatine phosphokinase (CPK) was only mildly elevated: 446 (normal range 26–308 U/L). Venous lactate level was initially elevated (4.5 on 6/16/14), but dropped to normal within 2 days (1.9 on 6/18/14; normal range: 0.4–2.0 mmoL/L). Chest x-ray on admission on 6/16/14 was clear.

Twenty-four hours after admission, the WBC count rose to 27.0 (from 8.0 on 6/16/14), and the differential band forms to 17% (from 1.3% on 6/16/14).

The right leg was first surgically debrided on 6/19/14, 3 days after admission. The tissue culture obtained at that time was negative. A photograph taken 3 days later on 6/22/14 is shown in Fig. 7.2b.

Table 7.2a General chemistry data noted on the day of admission – June 16, 2014 at 8.25am.

Parameter measured	Reference range	Measured value
Sodium	135–147 mmoL/L	140
Potassium	3.5–5.1 mmoL/L	4.0
Chloride	98–107 mmoL/L	107
Carbon dioxide	21–32 mmoL/L	27
Anion gap	3.0–11.0 mmol/L	6.0
BUN	7–18 mg/dL	15
Serum creatinine	0.6–1.3 mg/dL	1.0
eGFR MDRD:		
Non-Af Amer	>=60 mL/min/1.73 m^2	>60
Af Amer	>=60 mL/min/1.73 m^2	>60
Glucose	65–99 mg/dL	96
Calcium	8.5–10.1 mg/dL	10.0
Alkaline phosphatase	45–117 U/L	107
Albumin	3.4–5.0 g/dL	3.8
Total protein	6.4–8.2 g/dL	7.2
AST	15–37 U/L	20
ALT	12–78 U/L	44
Total bilirubin	0.2–1.0 mg/dL	0.8

Af Amer, African American; ALT, alanine aminotransferase; AST, aspartate aminotransferase; BUN, blood urea nitrogen; eGFR, estimated glomerular filtration rate; MDRD, calculated estimate of GFR; Non-Af Amer, non-African American.

Table 7.2b Complete blood count with differentials noted on the day of admission – June 16, 2014 at 8.25am.

Parameter measured	Reference range	Measured value
White blood cells	4.4–10.1 10*3/μL	8.0
Red blood cells	4.04–5.41 10*6/μL	5.01
Hemoglobin	11.9–16.8 g/dL	15.2
Hematocrit	36.3–50.6%	45.9
Mean corpuscular volume	83.3–100.5 fL	91.6
MCH	27.6–33.5 pg	30.3
MCHC	31.2–34.9 g/dL	33.1
RDW	38.5–54.4 fL	43.7
Platelets	117–369 10*3/μL	213
MPV	9.4–13.2 fL	9.4
Immature platelet fraction	1.0–7.0%	0.0 (L)
Neutrophils	43.70–84.90%	85.00 (H)
Lymphocytes	8.40–40.70%	12.10
Monocytes	3.40–12.00%	1.70 (L)
Eosinophils	0.60–6.00%	0.60
Basophils	0.20–1.20%	0.20
Immature granulocyte	<=0.5 %	0.4

MCH, mean corpuscular hemoglobin; MCHC, mean corpuscular hemoglobin concentration; MPV, mean platelet volume; RDW, red blood cell distribution width.

Table 7.2c Antimicrobial susceptibility of gram-negative organism from blood on June 16, 2014.

Antibiotic	MIC	Interpretation (sensitive/resistant)
Amikin	<16 μg/mL	Sensitive
Ampicillin	<8 μg/mL	Sensitive
Cefepime	<8 μg/mL	Sensitive
Cipro	<1 μg/mL	Sensitive
Claforan	<2 μg/mL	Sensitive
Fortaz	<1 μg/mL	Sensitive
Gentamicin	<4 μg/mL	Sensitive
Levaquin	<2 μg/mL	Sensitive
Meropenem	<1 μg/mL	Sensitive
Pip/Tazo	<16 μg/mL	Sensitive
Pipracill	<16 μg/mL	Sensitive
Primaxin	<1 μg/mL	Sensitive
Septra	<2/38 μg/mL	Sensitive
Tetracycline	<4 μg/mL	Sensitive
Unasyn	<8/4 μg/mL	Sensitive
Zinacef	<4 μg/mL	Sensitive

MIC, minimal inhibitory concentration.

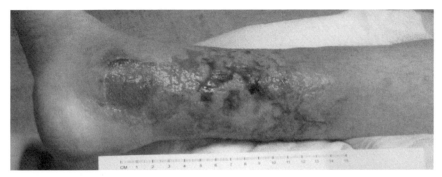

Figure 7.2b Progressive changes in the right leg over time and following initial surgical debridement. Photo taken on 6/22/14, 3 days after initial debridement on 6/19/14 (reproduced with permission).

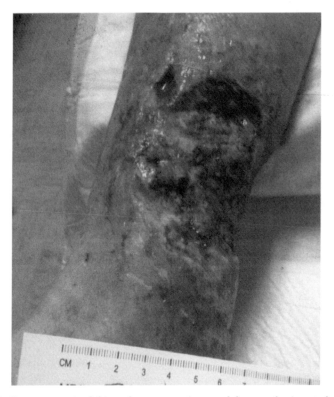

Figure 7.2c Severe necrosis of skin, subcutaneous tissue and down to fascia noted at right leg debridement on 6/25/14. Size of wound post debridement: 10 × 15 × 2 cm. Photo taken on 6/26/14 (reproduced with permission).

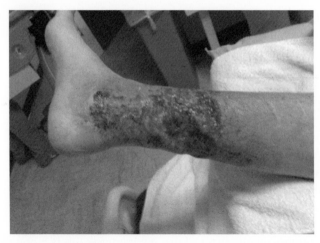

Figure 7.2d Severe necrosis of skin, subcutaneous tissue and down to fascia noted on right leg post second debridement on 6/25/14. Photo taken on 6/30/14, 5 days post second debridement (reproduced with permission).

Progression of the right leg wound over the next few weeks is shown in Figs. 7.2c and 7.2d. The patient was followed at the outpatient wound care center for several weeks. Split-thickness skin graft of the right leg wound was done on 8/20/14.

- *What is your final diagnosis? What is the likely organism causing this infection?*

The final diagnosis was bacteremia and right leg wound infection, due to *Vibrio* species. *Vibrio* species was due to or associated with necrotizing fasciitis and cellulitis of the right leg. The infection resulted from laceration of the right leg, the source of entry of the organism. Although the *Vibrio* was not speciated, it was most likely *Vibrio vulnificus*.

Comments and discussion

The rapidity of symptom onset in this patient after the leg laceration (18 hours) was a vital clue to the diagnosis. Blood cultures were reported positive within 11 hours after blood draw. The gram-negative organism (oxidase positive) grew well in standard media, typical for *Vibrio* organisms. The *Vibrio* species isolated in blood and wound was not resistant, but sensitive to all the agents tested. The patient received various antimicrobials, including ceftriaxone and doxycycline, to which the organism was sensitive.

This patient's admission was prolonged, from 6/16 to 6/30/14 (14 days). He was continued on outpatient wound care follow-up for several weeks. Outpatient treatments included UNNA compression wrap of the leg wound as well as wound debridement. Antimicrobial therapy was completed within 2–3 weeks after hospital discharge . He had a skin graft for wound closure on 8/20/14, about 2 months after the initial admission.

Typically, *Vibrio* species isolated in our hospital area over the past 15 years have been found to be sensitive to most of the usually available antimicrobials. High resistance would be very unusual. The *Vibrio* organism isolated in this patient was sensitive to most of the agents tested.

In this patient there was some delay in the initiation of aggressive surgical debridement (3 days) and some delay in the timing of the infectious disease consultation.

Lessons learned from this case

General measures that should be applied to patients with *Vibrio* sepsis or necrotizing fasciitis would include the following.

- Blood and wound cultures should be done as soon as possible after admission: these were done in this case.
- Rapid institution of antimicrobial therapy is crucial: this was done.
- Early aggressive surgical debridement should be instituted, and done as often as needed: there was delay in the initial aggressive debridement.
- Fluid resuscitation, metabolic, and other support measures should be pursued.
- Aggressive wound care and debridement as needed: prolonged outpatient wound care is typical, as was the case here.
- Some patients may need hyperbaric oxygen (HBO) treatment (this patient did not have HBO; however, this was not as important as aggressive surgical debridement).
- Most patients with severe necrotizing fasciitis would need skin grafting, typically weeks after admission, as was the case in this patient.

Further reading

Morris JG. *Vibrio vulnificus* infections. Available at: www.uptodate.com/contents/vibrio-vulnificus-infections?source=preview&anchor=H3115889#H3115889, accessed February 29, 2016.

Vibrio Illness (Vibriosis). Available at: www.cdc.gov/vibrio/index.html, accessed February 29, 2016.

Vibrio Illness (Vibriosis): *Vibrio vulnificus*. Available at: www.cdc.gov/vibrio/vibriov.html, accessed February 29, 2016.

Classification and characteristics of infectious diseases associated with trauma and outdoor activities

As previously noted, infectious diseases associated with trauma and outdoor activities are varied, but usually involve a breach of the skin or mucous membrane.

A summary of the characteristics and classification of infectious diseases associated with trauma and outdoors activities is given in Table 7.3a. The type of

Table 7.3a Characterization and classification of infectious diseases associated with trauma and outdoor activities.

Type of injury or mode of infection	Clinical syndromes	Common etiologic agents	Typical anatomic sites involved	Additional comments
Soft tissue	Cellulitis, myonecrosis, necrotizing fasciitis, sepsis	GAS, GCS, MSSA, MRSA, Clostridium, mixed anaerobic, Vibrio	Extremities, bloodstream, other body sites	These can range from a mild illness to sepsis with high mortality
Water related	Cellulitis, wound, gastroenteritis, sepsis	Aeromonas, Vibrio, M. marinum, Edwarsiella, Pleisiomonas	Extremities, GIT, bloodstream, skin	Some are indolent (Mycobacterium), others more fulminant (Vibrio vulnificus)
Penetrating/ puncture	Cellulitis, deep abscesses, gas gangrene	Strep., Staph., mixed anaerobic, GNB, Clostridium, sporotrichosis, other mold	Extremities, any organ or site	Clinical syndrome will depend on organ affected or involved, and the etiologic agent
Soil contamination	Cellulitis, abscesses	GNB, mold, Clostridium	Skin and soft tissue, any body site	To reduce complications, avoid primary closure of wounds
Gunshot wound or battlefield injury	Crush injuries, cellulitis, deep organ damage	GNB (Acinetobacter), mixed organisms, secondary infections	Extremities, head, any body site	Control of bleeding, amputation may save life, speed of access to good care essential to survival
Surgical wounds	Postoperative wound infection	MRSA, yeast, necrotizing fasciitis	Any anatomic site	Typically resistant organisms
Bites	Cellulitis, osteomyelitis, synovitis	Pasteurella, mixed agents, Capnocytophaga, Strep., Staph., GNB, anaerobic	Extremities, other soft tissue	Animal (pets: cat, dog; wild) and human bites. Tetanus and rabies vaccines may be needed
Burns	Wound infection (1–3° burns), sepsis	GPO, GNB, yeast, mold, often in that order	Any body site	Bloodstream infections; complications increase with larger (3°) burns and extended hospital stay
Injection drug use	Cellulitis, endocarditis	MSSA, Strep. (GAS, GCS), GNB, other agents	Extremities, blood, heart valves	Expect usual and unusual organisms. Younger patients are typically affected
Fish or crustacean associated	Fish tank granuloma, cellulitis, necrotizing fasciitis, sepsis	Mycobacterium marinum, Vibrio spp	Extremities (esp. UEs) for Mycobacterium; any extremity for Vibrio	Laceration, puncture from hook, barnacle, fish or crustacean. Avoid eating raw oysters

GAS, group A Streptococcus (Streptococcus pyogenes); GCS, group C Streptococcus; GIT, gastrointestinal tract; GNB, gram-negative bacteria; GPO, gram-positive organism; MRSA, methicillin-resistant Staphylococcus aureus; MSSA, methicillin-sensitive Staphylococcus aureus; UE, upper extremity.

injury, clinical syndrome, common etiologic agents, and typical anatomic sites involved are shown.

A common classification of surgical wounds is based on the level of expected bacterial contamination [1].

- Class I – clean wound
- Class II – clean-contaminated
- Class III – contaminated wound
- Class IV – dirty-infected wound.

A typical Class I (clean wound) is an uninfected operative wound in which no inflammation is encountered and the respiratory, alimentary, genital, or urinary tract is not entered. An example of a Class II wound is an operative wound in which the respiratory, alimentary, genital, or urinary tract is entered under controlled conditions and without unusual contamination. A Class III wound would include an open, fresh, accidental wound while a Class IV or dirty-contaminated wound typically includes old traumatic wounds with retained devitalized tissue and those that involve existing clinical infection or perforated viscera [1].

The human skin contains normal microbial flora, also called resident flora, as well as transient flora. The normal or resident flora is relatively fixed and would promptly re-establish itself if disturbed, while the transient or exogenous flora is usually of little significance, as long as the more permanent resident flora remains intact [2]. The location of the transient and resident bacterial flora on the skin is crucial in understanding the principles of antimicrobial prophylaxis and antisepsis [1].

Because of the variation of the "normal" microbial flora in different parts of the body [2], the site of surgery influences the choice of antimicrobial prophylaxis recommended before surgery or perioperatively. Mackowiak has summarized some of the predominant micro-organisms inhabiting various surfaces of the healthy human body [2]. Knowledge of the normal flora is useful in predicting the likely cause of an infection from a given site [2,3]. For example, the skin is colonized predominantly by gram-positive organisms while the upper intestine has relatively low numbers of micro-organisms (10^3–10^5/g content) compared to the large intestine, which has a preponderance of anaerobic organisms and up to 10^{10}–10^{11}/g stool [2–4]. I had occasion many years ago to review the concept of normal flora, and some of the factors affecting colonization resistance and infection with bacteria and *Candida* [3]. The resident flora provides some protection against invading micro-organisms, although this protection has limitations [3].

Figure 7.3a shows the location of transient and resident bacteria on the skin. The transient bacteria on the skin surface are easily removed, while the deep resident bacteria are not easily destroyed by sweating, bathing, or even skin antiseptics [1,3].

Over the past decades, procedures have been developed to try to diminish the risk of surgical wound infections. These include maneuvers to diminish

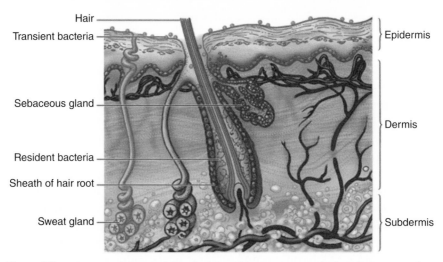

Hair

Transient bacteria

Epidermis

Sebaceous gland

Dermis

Resident bacteria

Sheath of hair root

Sweat gland

Subdermis

Figure 7.3a Schematic diagram of skin depicting skin structures, including the location of transient and resident bacteria. Source: From Talbot [1].

inoculation of bacteria into wounds that consider preoperative, intraoperative, and postoperative factors [1]. Such measures would include, for example, minimizing preoperative hospitalization, treating remote sites of infection, avoiding shaving or razor use at operative sites, ensuring timely administration and appropriate dosing of prophylactic antibiotics, careful skin preparation with the appropriate disinfectants, rigorous adherence to aseptic techniques, restriction of traffic in the operating room, maintenance of high flow of filtered air, etc.

Similarly, there are maneuvers that help to improve the host's ability to contain contaminating bacteria [1]. Some of these include preoperatively addressing malnutrition and obesity issues, discontinuing smoking and maximizing diabetes control, and intra- and postoperatively addressing factors to minimize dead space, devitalized tissue, and hematomas; maintenance of perioperative normothermia, and adequate hydration and nutrition.

These principles have been applied in the choice, timing, dosing and frequency of dosing of antimicrobial agents before and during surgery [1]. The antibiotic concentration in tissue over time following administration is crucial in the timing and dosing of perioperative antibiotics (antibiotic prophylaxis), designed to prevent surgical wound infections.

General comments about burns and bite injuries, including the epidemiology and pathogenesis of infections associated with these injuries and the typical resulting complications, have been addressed by Murray [5] and Goldstein [6].

In a survey of hospitalized burn patients in the 1980s, we found that gram-positive bacteria were the earliest skin colonizers (within the first week).

However, it was the enteric gram-negative organisms appearing in the second week that most frequently produced bacteremia, at about the same time that they were isolated from the burn wound [7].

Many of the improvements in morbidity and mortality associated with burns have been attributed to advances in burn shock resuscitation, airway management and ventilatory strategies, nutritional support, burn wound care, and infection control practices [5].

There appear to be some differences in the type of organisms seen in infections related to war trauma, compared to those seen in the civilian hospital environment. There is a continuing evolution in the type and nature of micro-organisms seen in previous wars, and the pattern noted since the Vietnam war as more gram-negative organisms than gram-positive organisms have been seen. Although the pathophysiologic explanation of this phenomenon is complex, the development of better surgical and antimicrobial treatments, including improved infection control, plus rapid evacuation of the patient from the site of war injury have clearly contributed to the improved survival of wounded soldiers, and likely influenced the nature of micro-organisms seen in such wound infections.

In a retrospective study of trauma casualties evacuated from the Iraqi war theater to the hospital ship USNS Comfort in 2003, the authors found that wound infections were more common than bloodstream infections (83% vs 38%) [8]. They also noted that gram-negative organisms were more common than gram-positive organisms. *Acinetobacter* species (36%) were the predominant organisms followed by *Escherichia coli* and *Pseudomonas* species (14% each) [8].

War wounds are distinct from peacetime traumatic injuries because of the high-velocity projectiles or blasts causing the injuries, and the consequent contamination of the complex and ragged wounds with clothing, soil or dirt, and environmental debris [8]. In a related study, some of the same authors noted that *Acinetobacter baumannii*-associated skin and soft tissue infection was an emerging infection in patients with traumatic war injuries [9].

In Table 7.3a, we present a summary of the characteristics of the various types of injury or mode of acquisition of infection following trauma or injury; the common clinical syndromes; common etiologic agents causing infection; as well as the typical anatomic sites involved in those infections. The list is obviously incomplete and limited, but the table is designed only as an example of how to categorize or classify trauma-related infections caused in a variety of conditions, and by myriad micro-organisms that can come from the patient's own normal flora (resident or transient: skin, mucosa, GIT), the environment of the injury (water, soil, clothing), inanimate objects (penetrating injuries from gunshot, other weaponry or blast), trauma (including burns and bite injuries), and iatrogenic causes (surgery, etc.).

Prevention and management of these infections require an understanding of the concept of body site-specific normal microbial flora [3]; the need for intervention measures to diminish bacterial contamination of wounds (preoperative, intraoperative, and postoperative maneuvers); improvement in the patient's host factors preoperatively, perioperatively, and postoperatively; appropriate antimicrobial prophylaxis, with respect to choice of agent, timing, duration and dosing of antimicrobial agents for the site, and nature of the surgery or infection of concern. Good infection prevention and control practices are also required, as well as application of the modern technology of critical care medicine.

References

1 Talbot TR. Surgical site infections and antimicrobial prophylaxis. In: Mandell GL, Bennett JE, Dolin R (eds) *Principles and Practice of Infectious Diseases*, 7th edn. Philadelphia: Churchill Livingstone, 2010, pp. 3891–904.

2 Mackowiak PA. The normal microbial flora. New England Journal of Medicine 1982; 307: 83–93.

3 Ekenna O. Normal flora, colonization resistance, and infection with bacteria and Candida. Nigerian Medical Journal 1992; 23: 47–54.

4 Gorbach SL, Plaut AG, Nahas L, Weinstein L, Spanknebel G, Levitan R. Microorganisms of the small intestine and their relations to oral and fecal flora. Gastroenterology 1967; 53: 856–67.

5 Murray CK. Burns. In: Mandell GL, Bennett JE, Dolin R (eds) *Principles and Practice of Infectious Diseases*, 7th edn. Philadelphia: Churchill Livingstone, 2010, pp. 3905–9.

6 Goldstein EJC. Bites. In: Mandell GL, Bennett JE, Dolin R (eds) *Principles and Practice of Infectious Diseases*, 7th edn. Philadelphia: Churchill Livingstone, 2010, pp. 3911–15.

7 Ekenna O, Sherertz RJ, Bingham H. Natural history of bloodstream infections in a burn patient population: the importance of candidemia. American Journal of Infection Control 1993; 21: 189–95.

8 Petersen K, Riddle MS, Danko JR, et al. Trauma-related infections in battlefield casualties from Iraq. Annals of Surgery 2007; 245(5): 803–11. Available at: www.ncbi.nlm.nih.gov/pmc/articles/PMC1877069/, accessed March 1, 2016.

9 Sebeny PJ, Riddle MS, Petersen K. *Acinetobacter baumannii* skin and soft-tissue infection associated with war trauma. Clinical Infectious Diseases 2008 ; 47(4): 444–9. Available at: http://cid.oxfordjournals.org/content/47/4/444.short, accessed March 1, 2016.

Further reading

Management of *Vibrio vulnificus* wound infections after a disaster. Available at: http://emergency.cdc.gov/disasters/disease/vibriofaq.asp, accessed March 1, 2016.

Vibrio illnesses after Hurricane Katrina – multiple states, August–September 2005. Morbidity and Mortality Weekly Report 2005; 54(37): 928–31.

Griffith GE. Pathogenesis of nontuberculous mycobacterial infections. Available at: www.uptodate.com/contents/pathogenesis-of-nontuberculous-mycobacterial-infections?topicKey=ID%2F5345&elapsedTimeMs=4&source=search_result&selectedTitle=9%7E150&view=print&displayedView=full, accessed March 1, 2016.

Morris JG. *Vibrio vulnificus* infections. Available at: www.uptodate.com/contents/vibrio-vulnificus-infections?source=preview&anchor=H3115889#H3115889, accessed March 1, 2016.

Morris JG, Horneman A. Aeromonas infections. Available at: www.uptodate.com/contents/aeromonas-infections?topicKey=ID%2F3138&elapsedTimeMs=5&source=search_result&searchTerm=Aeromonas+infections&selectedTitle=1%7E28&view=print&displayedView=full, accessed March 1, 2016.

CHAPTER 8

Acute and Chronic Subcutaneous Fungal Infections

Case 8.1 Exophytic skin lesion masquerading as cancer

An 83-year-old Caucasian female was referred in November, 2004 for evaluation and treatment following excisional biopsy of a left hand dorsum lesion done 2 weeks earlier. The histopathology had suggested the presence of blastomycosis. Three months earlier in August, 2004, a previous biopsy of the same lesion had been described as a well-differentiated squamous cell carcinoma, with keratoacanthoma pattern, in an outside institution.

The past medical history was significant for severe chronic obstructive pulmonary disease (COPD) on home oxygen, chronic steroid therapy for the COPD, chronic congestive heart failure (CHF), atrial fibrillation, coronary artery disease (CAD) with previous three-vessel coronary artery bypass graft (CABG) surgery in 1995. Furthermore, she had ecchymotic skin bruising in the upper and lower extremities related to chronic prednisone therapy.

Epidemiologic history and review of systems suggested the presence of the hand dorsum lesion for at least 6 months. It had been indolent and slowly growing. The lesion was not painful, but there was occasional itch. She had done some gardening and worked with plants, but there was no obvious history of observed skin puncture injury recently. She had, however, a history of frequent falls (six times in the previous summer), and had been staying with her son following her most recent fall 2 months earlier. She had fragile skin thought to be related to steroid therapy for several years for the COPD. She had no fever. She lived in a small town in the southern United States, on the Gulf Coast.

She was a retired nurse who was widowed 4 years earlier. She started smoking rather late in life, and had smoked one pack of cigarettes/day for 25 years at presentation. She did not drink alcohol. She was allergic to penicillin (nausea).

Medications included ipratropium bromide, albuterol, digoxin, furosemide, potassium chloride, children's aspirin, and prednisone 20 mg daily.

Cases in Clinical Infectious Disease Practice: Obtaining a Good History from the Patient Remains the Cornerstone of an Accurate Clinical Diagnosis: Lessons Learned in Many Years of Clinical Practice,
First Edition. Okechukwu Ekenna.
© 2016 John Wiley & Sons, Inc. Published 2016 by John Wiley & Sons, Inc.

The initial examination (in October, 2004) by the plastic surgeon referring the patient suggested no pain, fever, or palpable lymphadenopathy in the axilla or clavicular areas.

The exam in the office showed her to be in no acute distress, but chronically ill looking. Vital signs were as follows: blood pressure (BP) 138/78, respiratory rate (RR) 18, heart rate (HR) 78, temperature 96°F; height 5′4″, weight 130 pounds. The head and neck exam showed no thrush, but loss of most of her teeth. Heart, lung, and abdominal exams were unremarkable for her age; however, the extremities showed multiple ecchymotic skin changes consistent with fragile and atrophic skin of old age and chronic steroid therapy. The neurologic exam showed no focal findings, but the patient was in a wheelchair for more comfort and fall prevention. The site of the biopsy on the left hand dorsum showed atrophic skin. Photographs of the hand are shown in Fig. 8.1a, including the verrucose and keratotic hand lesion as seen prior to the excisional biopsy.

Follow-up course

On the assumption that this was indeed cutaneous blastomycosis, the surgical excision performed prior to the office visit would be an appropriate therapy if all the lesions had been excised. Additional therapy with itraconazole or amphotericin B would mean extended treatment for up to 6 months, with the potential for liver or renal toxicities in this elderly patient. We opted, therefore, to watch the patient closely and follow her for evidence for recurrence, and re-biopsy her at that time for fungal culture (not done with the previous biopsies), along with histopathology. We would be particularly interested in any satellite lesions that showed up.

On 2/22/05, the patient was admitted to hospital with symptoms of worsening shortness of breath (SOB), confirmed to be due to exacerbation of COPD. In

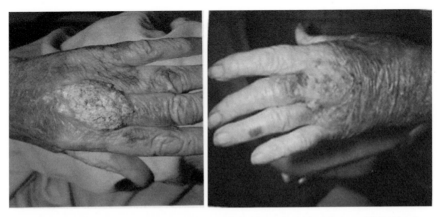

Figure 8.1a Left-hand photo taken on 10/12/04; photo on the right taken on 11/3/04 (after excision of hand dorsum lesion) (reproduced with permission).

addition, she was found to have a left ventricular ejection fraction of 25–30%, consistent with CHF.

While these medical conditions were being addressed, infectious diseases consult was requested on 2/28/05 to address a possible recurrence of the hand lesion excised 3 months earlier.

On examination, she was thin and chronically ill. She had lost weight (now 104 pounds), was afebrile, and had thrush. Her vital signs were otherwise stable. The hand exam showed a small but visible new growth of tissue on the dorsum of the left third finger, just proximal to the proximal interphalangeal (PIP) joint (Fig. 8.1b).

Biopsy of the lesion on the third finger was recommended. This procedure was carried out 3 days later on 3/3/05. The histopathology showed skin and subcutaneous tissue with hyperkeratosis and pseudoepitheliomatous hyperplasia. Within the dermis, there was microabscess formation with occasional giant cells. Also, numerous budding yeasts and septate, thick-walled hyphae were identified. A preliminary diagnosis was made based on review of the histopathology.

• *What is your diagnosis?*

The biopsy specimen was sent for special stains and culture, including fungal culture. Acid-fast bacilli (AFB) and KOH stains were negative. Baseline liver function tests were done, and the patient was started on itraconazole (200 mg daily) by mouth. She was hospitalized from 2/22/05 to 3/8/05, and subsequently discharged to a skilled nursing facility (nursing home) for supportive care.

The mycobacterial cultures were subsequently negative. She had many other more serious underlying diseases.

Four months later, on July 5, 2005, she was still at the nursing home. She was unable to show up for any outpatient follow-up visits. The final outcome on this patient is therefore unknown.

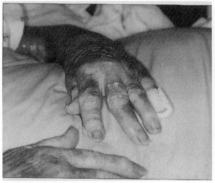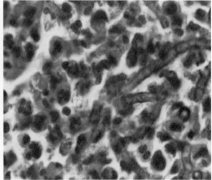

Figure 8.1b New keratotic lesion on third left finger (photo taken 2/28/05), and histopathology of the biopsy (H&E stain) done on 3/3/05. Note the giant cells and the large brown, thick-walled, septate hyphae (reproduced with permission). (*See insert for color information*)

The working diagnosis was chromoblastomycosis (chromomycosis). This is defined as a chronic, cutaneous, and subcutaneous opportunistic infection characterized by the presence of brown-pigmented (dematiaceous), rounded sclerotic bodies in infected tissues.

- *The culture was ultimately confirmed to be Madurella grisea (one of several agents of chromomycosis).*

The characteristics of the dematiaceous fungi causing chromoblastomycosis will be discussed after the next case.

Case 8.2 Chronic persistent subcutaneous fungal infection of the leg

A 72-year-old black female was referred in June, 2001 for chronic persistent exophytic skin lesions in the right leg. Two weeks earlier, she had seen a dermatologist who did a biopsy of this leg lesion, and then referred her for further evaluation.

The initial observation was a small non-painful bump in her fourth toe on the right foot noted about 8 years previously. Although these lesions bruised easily, they were not painful, except when bumped. She tried over-the-counter topical antimicrobial treatments without benefit. She was able to live with it, however.

One and a half years before the office visit (about December, 2000), she noted new lesions on her right ankle that spread to involve most of the lateral ankle and some of the posterior ankle, within months. These were multiple, bumpy, and nodular lesions that failed to respond to various topical ointments, including antibacterial and antifungal regimens, as well as oral systemic antibiotics. The primary care physician then referred her to the dermatologist for a biopsy.

As a young woman, she grew up on a farm in South Alabama where she spent a lot of time planting crops and raising hogs and other animals. She later moved to South Mississippi, and had a garden where she grew a variety of vegetables.

She had no associated fever or pain. Occasionally, there would be minimal bleeding when she bumped her affected toe or ankle, but no purulent discharge or drainage was noted. There was only minimal swelling of the ankle, and her activities of daily living, including working as a cook, were not adversely affected. She had no systemic complaints.

Her past medical history was significant for hypothyroidism, hypertension, type 2 diabetes mellitus, osteoarthritis, previous cataract surgeries, and gastroesophageal reflux disease. She lived with her husband, and had many grown children.

Her medications prior to the office visit included butalbital/acetaminophen/caffeine for headaches, verapamil, omeprazole, pancrelipase, levothyroxine, metformin, cephalexin, and terbinafine.

On examination, she was alert and oriented, and not in acute distress. Her vital signs were as follows: BP 128/78, RR 20/minute, HR 88, temperature 99.1°F; height 5′4″, weight 198 pounds. The head and neck exam showed prosthetic dentures, but was otherwise unremarkable. The heart, lung, and abdominal exam showed nothing significant, and no adenopathy was noted in the neck, axilla, or inguinal areas. The upper extremities were unremarkable while the right ankle and foot showed nodular and indurated lesions in the fourth toe and the lateral and posterior ankle (Fig. 8.2a). The right ankle showed non-pitting puffiness. Neurologic exam was normal.

- *What are your initial differential diagnoses? How would you proceed to make or confirm the diagnosis?*

A biopsy of one of the right ankle lesions was done by the dermatologist 2 weeks prior to the office infectious diseases consult, but no cultures were done. A repeat biopsy of the skin nodule was scheduled, along with accompanying fungal cultures.

Laboratory findings

The histopathology on 6/12/01 showed a dermis with extensive infiltration of polymorphous granulation tissue with many multinucleated giant cells and small abscesses composed of neutrophils. On H&E and fungal stains, spores were present. The spores were brown in the H&E sections. The stains were similar on repeat biopsy on 7/2/01.

A working (presumptive) diagnosis of chromo (blasto) mycosis (a dematiaceous fungal infection) was made, based on the clinical history, physical examination, and histopathologic features noted in the biopsy.

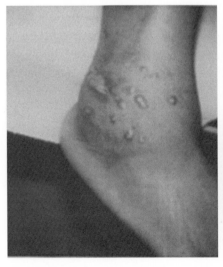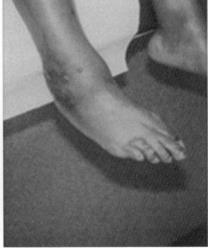

Figure 8.2a Right ankle and foot: photo taken on 6/27/01 (reproduced with permission).

The patient was started empirically on oral antifungal medication after the punch biopsy was done, while awaiting the results of the cultures, expected to be at least several weeks in the making.

Outpatient follow-up course

Acid-fast bacilli (AFB) stains on the tissue biopsy were negative, as well as final AFB cultures weeks later. KOH prep showed "fungal elements." Initial growth on the fungal culture medium was, however, noted on day 17; the growth was then subcultured to Sabouraud dextrose agar (SDA) medium (Fig. 8.2b). On day 20 of the new growth, a final diagnosis was made, based on macroscopic and microscopic characteristics. The mold was identified as *Fonsecaea pedrosoi.*

Fungal susceptibility testing was done at a reference fungus testing laboratory at the University of Texas in San Antonio, Texas. The organism was reconfirmed, and found to be susceptible to the following antifungal agents.

- Itraconazole, with minimum inhibitory concentration (MIC) (µg/mL) of <0.015 at 72 and 96 hours
- Voriconazole, with MIC (µg/mL) of <0.125 at 72 and 96 hours
- Posaconazole, with MIC (µg/mL) of <0.015 at 72 and 96 hours

At the time of the antifungal testing (August, 2001), the last two agents were either new or not fully licensed for general use.

The patient had been started on itraconazole, at 200 mg daily by mouth, after the skin punch biopsy of 7/2/01. Treatment was continued during the subsequent follow-up.

The patient was seen in the office initially every few weeks till a formal cultural diagnosis was made, and thereafter every few months. She remained stable but would sometimes relapse, with drainage, itching or spread of the skin nodules in the right leg. The lesions never spread beyond the area below the knee on the right.

Beginning in 2005, the dermatologist tried several additional treatments with cryotherapy, with shaving off of some of the skin lesions on the leg, leading to flattening of the lesions. Definitive curative surgery was not possible because that

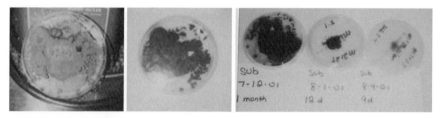

Figure 8.2b A 33-day-old culture on SDA (top). Velvety, powdery black mold culture (middle) is a close-up of the powdery mold on the left; right shows bottom of agar with growth progression of mold from days 9, 12, and 33. Photo taken on 8/13/01.

would have required amputation, which was unacceptable. She was not always able to afford the cost of itraconazole therapy, and took that medication after a few years only intermittently.

Her clinical course was that of a slow but progressive disease, with only very limited response to antifungal therapy. Recurrences in the form of new lesions or abscesses appeared when treatment was interrupted. Her liver function tests, however, remained normal throughout her 12 years of follow-up. The last office visit was in 2013.

At 85 years of age, she has other underlying diseases, such as peripheral vascular disease, to worry about. She has finally accepted that there can be no cure for this chronic subcutaneous fungal infection, persistent and progressive over the last 20 years. The most recent office photos were in 2009 and 2013 (Figs 8.2c, 8.2d).

• *Final diagnosis: Fonsecaea pedrosoi, a dematiaceous, chronic subcutaneous mycosis.*

Case discussion and characteristics of chromomycosis

The subcutaneous mycoses are generally rare fungal infections that cause a variety of acute, subacute, and chronic infections that range from debilitating

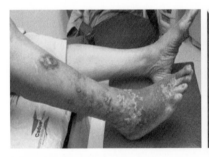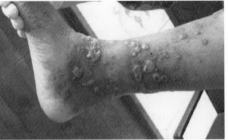

Figure 8.2c Progressive chromomycosis of the right leg, 16 years after onset. Photo taken on 5/29/09 (reproduced with permission). (*See insert for color information*)

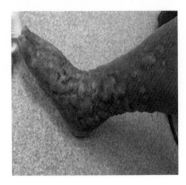

Figure 8.2d Progressive chromomycosis of the right leg, 20 years after onset. Photo taken on 6/28/13 (reproduced with permission).

Table 8.2a The subcutaneous mycoses.

Disease entity	Causative agents
Sporotrichosis	*Sporothrix schenckii*
Chromoblastomycosis	*Fonsecaea, Phialophora, Cladosporium*
Phaeohyphomycosis	*Cladosporium, Exophiala, Wangiella, Bipolaris, Exserohilium, Curvularia*
Mycotic mycetoma	*Pseudoallerscheria, Madurella, Acremonium, Exophiala*
Subcutaneous zygomycosis (entomophthoromycosis)	*Basidiobolus ranarum, Conidiobolus coronatus*
Subcutaneous zygomycosis (mucormycosis)	*Rhizopus, Mucor, Rhizomucor, Absidia, Saksenaea, Apophysomyces*
Rhinosporidiosis	*Rhinosporidium seeberi*
Lobomycosis	*Loboa loboi*

Adapted from Mycology Online (University of Adelaide, Australia). Available at: www.mycology.adelaide.edu.au/Mycoses/Subcutaneous/, accessed March 1, 2016.

to life-threatening infections in normal and compromised hosts. As shown in Table 8.2a, the causative agents or organisms vary, and usually produce specific or peculiar disease entities.

The patient described here had chromomycosis, also called chromoblastomycosis. This disease entity is described as a chronic, cutaneous, and subcutaneous opportunistic infection characterized by the presence of brown-pigmented (dematiaceous), rounded sclerotic bodies in infected tissue. This patient obviously had a very chronic infection that has persisted for over 20 years.

A characteristic of dematiaceous fungi is that the brown to black color in the walls of their vegetative cells (hyphae) and conidia, or both, results in colonies that range from olive or gray to black on visual inspection.

As seen in the patients reported here and earlier (*Madurella* patient), clinically, lesions most often appear on exposed parts of the body, usually the limbs. The lesions usually start as small scaly papules or nodules that are painless and may be itchy. Over time (years), however, they may become tumorous or cauliflower-like. Other features include epithelial hyperplasia, fibrosis, and microabscess formation in the epidermis, seen best on histopathologic stains.

Chromoblastomycosis must be differentiated from other cutaneous fungal infections such as blastomycosis, lobomycosis, paracoccidioidomycosis, and sporotrichosis (see Table 8.2a). They must also be differentiated from exophytic or verrucose lesions that include protothecosis, leishmaniasis, verrucose tuberculosis, certain leprous lesions and syphilis, and rarely squamous cell cancer.

The initial laboratory diagnosis is by skin scraping or biopsy of a nodule or lesion. The dematiaceous nature is usually obvious on histopathology. However, because sclerotic bodies seen in histopathology are similar in all agents of chromomycosis, the causative fungus can only be identified in culture, not by tissue morphology.

Preliminary diagnosis is made by skin scraping, using direct microscopy (KOH, Parker ink or calcofluor stain), or by tissue sections (hematoxylin and eosin [H&E], periodic acid-Schiff [PAS], or Grocott's methenamine silver [GMS]). Definitive diagnosis is by fungal culture using SDA followed by macroscopic and microscopic characteristics: color, growth rate, arrangement of conidia, etc.

The epidemiology of chromomycosis suggests that it is predominantly a disease of rural populations, especially in tropical and subtropical climates (30° N–30° S). The disease is chronic in nature (rare in children, more common in adults) and more common in barefooted agricultural populations. It is associated with decaying vegetation or soil.

Our patient spent a lot of her youth on a farm, and later gardening in the subtropical climate of the South. She had never traveled out of the southern United States before her illness. Infection usually results from traumatic cutaneous injury, often unrecognized (wood splinters or thorns). The previous patient with *Madurella grisea* infection was also a gardener.

Common etiologic agents of chromomycosis include *Phialophora verrucosa*, *Fonsecaea pedrosoi*, *F. compacta*, and *Cladosporium carrionii*.

Typical and common complications include ulceration, lymphedema, and secondary bacterial infection. Mortality is very rare. Our patient was alive and well after 20 years but has no prospect of a cure of this illness since curative surgery (amputation) was impractical.

The literature suggests that treatment is generally difficult and refractory. Successful surgical excision requires a wide margin to prevent local dissemination. Itraconazole 200–400 mg produces improvement but relapse is common. Complete cure is rarely reached in severe cases. Other agents can be useful: terbinafine (Lamisil); posaconazole (a new azole); 5FC plus itraconazole. Treatment may last for years. Heat (e.g. pocket warmers) has been found useful for management.

Lessons learned from this case

- A definitive diagnosis of chromomycosis (chromoblastomycosis) can only be made with culture and stain of lesion, biopsy, or scraping.
- Histopathology is very useful and rapid in determining the dematiaceous nature of the nodule, allowing for additional studies needed to arrive at a definitive diagnosis.
- Chromomycosis is usually a chronic, persistent subcutaneous infection of the limbs, the typical sites of organism inoculation.
- It is typically an indolent infection with low mortality.

- Surgical excision, if done early, may be curative. The site of infection and the duration and extent of the lesions may affect the choice of therapy.
- Antifungal therapy is rarely curative alone without surgical excision, unless the lesion is small and caught very early in the process.
- Multiple modalities of treatment may be required to adequately manage this illness.
- A high index of suspicion is required to make a diagnosis of this rare disease in the United States. A detailed history, including failed previous antimicrobial treatments, is part of the relevant epidemiology that should raise the index of suspicion.
- Most affected patients are adults (it takes a long time to develop), and usually have exposure to farming or gardening in the tropics or subtropical environment.

Further reading

Bustamante B, Campos PE. Eumycetoma. Available at: www.uptodate.com/contents/eumycetoma?view=print, accessed March 1, 2016.

Dixon DM, Polak-Wyss A. The medically important dematiaceous fungi and their identification. Mycosis 1991; 34(1-2): 1–18.

Hospenthal DR. Agents of chromomycosis. In: Mandell GL, Bennett JE, Dolin R (eds) *Principles and Practice of Infectious Diseases*, 7th edn. Philadelphia: Churchill Livingstone, 2010, pp. 3277–80.

Hospenthal DR. Agents of mycetoma. In: Mandell GL, Bennett JE, Dolin R (eds) *Principles and Practice of Infectious Diseases*, 7th edn. Philadelphia: Churchill Livingstone, 2010, pp. 3281–5.

Hospenthal DR. Uncommon fungi and prototheca: dark-walled fungi and agents of phaeohyphomycosis. In: Mandell GL, Bennett JE, Dolin R (eds) *Principles and Practice of Infectious Diseases*, 7th edn. Philadelphia: Churchill Livingstone, 2010, pp. 3365–76.

Case 8.3   A young man with a laceration injury contaminated with soil

A 19-year-old Caucasian male was admitted in June, 1997 following a laceration injury to the right calf suffered while working outdoors. The injury occurred when he fell into a ditch during construction work and the blade of a dirt-digging machine, a "ditch witch," injured him. He was promptly rescued and brought to the emergency room (ER) within 1 hour of the accident.

Following admission, the wound was cleansed and closed with a few staples. He was discharged in a stable condition 2 days later on oral cephalexin. No cultures were reportedly done.

Over the next several days he noted increasing tenderness, swelling, and non-foul-smelling drainage from the right calf, and so returned to the ER 8 days later for further evaluation, and was admitted.

His past medical history was generally unremarkable. His tetanus immunization was up to date. He smoked a pack of cigarettes a day, and did not use any recreational drugs or alcohol. He had no known drug allergies. He had been taking the prescribed cephalexin and pain medications.

The review of systems was positive for low-grade fever (temperature $\leq 100°$F) and a one-time chill. The main complaint was pain and tenderness in the right calf, the swelling, and some slight drainage at the site.

Initial hospital course

After admission on 7/6/97, he was placed on various antimicrobial agents, specifically, cefotaxime, gentamicin, and ofloxacin. Cultures obtained on admission showed a growth of *Aeromonas hydrophila*, sensitive to the quinolones and third-generation cephalosporins.

Subsequent cultures were obtained during the multiple surgical procedures that were performed on 7/10, 7/12, 7/14, 7/17, and 7/24/97. These surgical procedures (incision, drainage, irrigation, and debridement) were designed to keep the contaminated wound clean. On 7/10/97, *S. aureus* (methicillin resistant) and *S. epidermidis* were cultured from the right calf wound.

Infectious diseases consult was sought 9 days later on 7/15/97, especially because of the multiple organisms cultured from the leg wound.

On examination on 7/15/97, the patient was alert and oriented, and in no acute distress. His vital signs were as follows: BP 122/70, RR 24, HR 96, temperature 99.9°F; height approximately 6' and weight about 218 pounds. The head and neck, heart, lung, abdomen, and neurologic examinations were unremarkable. The skin showed no rash, and lymph nodes were not palpably enlarged in the neck, axilla, or groins. The only noted abnormality was the leg examination. The right leg (calf) dressing was removed, but the packing of the wound from the surgery the day before (7/14/97) was not removed. The left leg was unremarkable, except for an old, healed scar in the medial portion of the left big toe.

A photograph of the calf wound taken by the family, 3 days before readmission, is shown in Fig. 8.3a. The wound involved a good portion of the lateral calf, as seen in the picture.

Admission laboratory findings showed a WBC of 11.8, hemoglobin/hematocrit (H/H) of 14.8/43.3, respectively, and platelet count of 302,000. The differential count showed 74% polymorphs and 16% lymphocytes. General chemistry was unremarkable, while HIV and HCV serology were negative. The chest x-ray showed vague and diffuse interstitial changes, but no pneumonia.

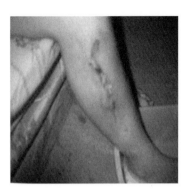

Figure 8.3a Photo taken by the patient's family on 7/3/97, 3 days before readmission on 7/6/97.

Following patient evaluation and review of the microbiology data with the laboratory staff, it became apparent that an unidentified "mold" had been noted on 7/15/97, from the 7/10/97 wound culture. This mold growth was noted on top of the back-up (thioglycholate) broth culture. Adjustments in antimicrobial therapy were made, and specific fungal stains and cultures were requested to be done during the next incision and drainage surgery. Antibacterial regimen was now tailored to cover the isolated *Aeromonas* and MRSA, as well as non-isolated agents, like *Actinomyces* and *Nocardia*. Trimethoprim/sulfamethoxazole and clindamycin were added, replacing gentamicin and cefotaxime, while ofloxacin was continued.

Follow-up hospital course and new laboratory data

On examination of the patient 2 days later on 7/17/97, the right calf area was noted to be edematous; the wound had necrotic black eschar edges and also had yellowish brownish exudate. The surgeon was encouraged to continue surgical wound debridement and whirlpool treatments, and advised not to close the wound for a while until all the necrotic material had been completely debrided and the wound was clean. Additional fungal stains and cultures were obtained at surgery on that day.

On 7/20/97, with the report of a hyaline mold noted in the wound cultures of 7/17/97, itraconazole was added to provide antifungal coverage, until more data was available.

Although this hyaline mold with rapid growth (2–3 days) appeared initially to have septate hyphae on tease mount, a preliminary report received on 7/22/97 on the previous culture of 7/10/97 (sent off on 7/15/97) suggested otherwise. This was indeed a non-septate, rapidly growing mold, suggestive of a zygomycete.

Amphotericin B was therefore started in place of itraconazole on 7/22/97. This treatment was subsequently continued for 2 months till 9/23/97, a total dose of 2100 mg. Initially, this regimen was given daily but later it was changed to three times a week. A secure central line (Hickman catheter) was placed before discharge from the hospital, about 3 weeks after admission. This was deemed necessary to safely deliver the amphotericin B therapy.

At the first outpatient office visit 1 month after the July admission (on 8/6/97), the antibiotics were discontinued, while the amphotericin B was continued. Clindamycin had been discontinued earlier, some 6 days before discharge from the hospital. The creatinine reached a peak of 4.0 mg/dL on 8/2/97, a few days after discharge from the hospital, but eventually normalized again 6–8 weeks after completion of therapy.

Laboratory findings on 11/25/97, 2 months after completion of therapy, showed a WBC of 6.1, H/H 13.5/39.1, respectively, MCV 86.8, platelets 186,000, and a normal differential count. Erythrocyte sedimentation rate (ESR) was normal at 9 mm/hour and general chemistry, including potassium (3.9), and liver function tests were all within normal limits. Creatinine was back to normal at 1.3 mg/dL.

The patient was followed closely during treatment of the wound infection, and was seen in the office after discharge from the hospital on 8/6, 8/12, 9/2, 9/18, 9/23, and 11/25/97, as well as on 7/9/98 for a 1-year final follow-up visit. The wound was finally surgically closed on 8/14/97.

Figure 8.3b shows photographs of the right leg wound taken during outpatient follow-up visits.

Twelve years later (in 2009), the patient was well and still working, and now was the father of several children.

- *What is the likely diagnosis in this patient?*
- *How would you classify this form of zygomycete or mucormycosis?*

Case discussion

This case illustrates several important issues in dealing with potentially contaminated wounds. First, this patient had an outdoor injury contaminated with soil, during the hot summer season of the Gulf Coast of Mississippi. Because the laceration was from the blade of a construction vehicle used in digging up dirt, it was likely that the infecting organisms were carried deep into the tissue, even if the patient had not fallen into the ditch in the first place.

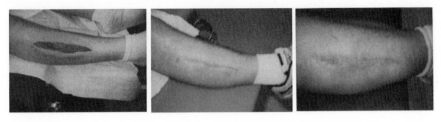

Figure 8.3b Progressive healing of right calf wound. Photos were taken on 8/6/97, 11/25/97, and 7/7/98.

During the initial admission of 6/27/97, the wound was washed out but no cultures were obtained. The first organism isolated during the second admission on 7/6/97, *Aeromonas hydrophila*, would not have responded to the cephalexin that the patient was treated with in the week before readmission, because it was (and usually is) resistant to first-generation cephalosporins. This organism and the isolated *Staphylococcus aureus* were adequately treated subsequently during the second admission. It is likely that both of these agents played a role in the symptomatology of the patient (wound drainage, edema, pain, and swelling of the calf), that lead to the readmission.

The more likely cause of the prolonged wound infection, however, was the mold (fungus) that contaminated the wound at the time of the injury. The fact that multiple wound cultures for fungus were found to be positive on several days (7/10, 7/12, and 7/17/97), with the same mold, supports the notion that this organism was a true pathogen.

Apophysomyces elegans was the fungus cultured from the right calf wound on multiple occasions as noted above. The organism identification was reconfirmed when the isolate was sent to a reference fungus testing laboratory (University of Texas in San Antonio). The local laboratory that did the initial identification used the water culture method to induce sporulation, since this fungus does not readily sporulate on routine fungus media, following the procedures outlined by Larone [1].

Apophysomyces elegans is a thermophile organism that can be isolated from the soil, the likely source of this infection. It likely contaminated the wound during the laceration injury. It grows rapidly, typically within 4 days, at temperatures up to 42°C. The fluffy, cottony growth (often white to cream) fills up the plate in several days. The special culture method for sporulation, and the differentiation from similar zygomycetes, like *Absidia*, have been outlined by Larone [1] and others [2].

Infections due to zygomycetes are usually divided into six clinical syndromes: rhinocerebral, pulmonary, disseminated, gastrointestinal, cutaneous, and unusual or miscellaneous [3]. Our patient had the cutaneous form of the disease, acquired through inoculation of the organism from the soil.

A review of the literature shows that infection with *A. elegans* is still rare, with most reported cases being typically case reports [4–7]. Most of these reported cases were in normal, immunocompetent hosts. However, in a study from India, Chakrabarti et al. reported that uncontrolled diabetes mellitus was found in 63% of patients studied (10/16) [8]. Only five of these 16 had cutaneous infection with *Apophysomyces*, three of whom also had diabetes mellitus. Mortality was 37.5% (6/16) in this study. All 16 patients had some form of surgery. Fourteen of 16 patients (88%) in these series were treated with a combination of surgery and amphotericin B therapy. The two patients who did not receive amphotericin B therapy died.

Successful treatment of patients with *Apophysomyces* infections requires aggressive and often multiple surgical debridement, as was done in our patient, as well as extended antifungal therapy.

Our patient had a very good outcome and recovered fully, without limb amputation. The expected treatment complications of normocytic anemia, leukopenia, hypokalemia, and renal function abnormality all returned to normal within 6–8 weeks of completion of 2100 mg amphotericin B therapy given over a 2-month period.

- *Final diagnosis: Apophysomyces elegans right calf wound infection acquired through inoculation of the organism through soil contamination during laceration injury.*

Lessons learned from this case

- A contaminated wound must be debrided and washed out as often as necessary to remove all debris and potential pathogens.
- There should never be a rush to close a contaminated wound prematurely.
- Repeated positive culture for "mold" is a sure sign that the wound is infected, and that the culture report does not reflect a skin or laboratory contaminant. When in doubt, repeat the fungal culture during subsequent wound debridement.
- The report of rapid growth of a hyaline mold that is non-septate should prompt one to think of mucormycosis or zygomycoses. This information is crucial for institution of appropriate and rapid therapy, and depends on the competence and expertise of the clinical microbiologist on the ground.
- Although the classification of the agents of mucormycosis and related diseases is complicated (and in flux), it is still crucial to separate septate from non-septate molds by simple microscopy (e.g. tease mounts). The non-septate molds must be identified and treated early, in order to achieve a good outcome for the patient.
- Combination therapy was needed for a good outcome (surgery, wound care, antifungal therapy), and patience to deal with the expected and reversible complications of therapy, in this case, renal insufficiency.

References

1 Larone, DH. Zygomycetes: *Apophysomyces elegans*. In: *Medically Important Fungi. A Guide to Identification*, 3rd edn. Washington, DC: ASM Press, 1995, pp. 112–113.
2 Richardson MD, Koukila-Kahkola P. Rhizopus, Rhizomucor, Absidia, and other agents of systemic and subcutaneous zygomycoses. In: Murray PR, Baron EJ, Jorgensen JH, Landry ML, Pfaller MA (eds) *Manual of Clinical Microbiology*, 9th edn. Washington, DC: ASM Press, 2007, pp. 1839–56.
3 Kontoyiannis DP, Lewis RE. Agents of mucormycosis and entomorphthoramycosis. In: Mandell GL, Bennett JE, Dolin R (eds) *Principles and Practice of Infectious Diseases*, 7th edn. Philadelphia: Churchill Livingstone, 2010, pp. 3257–69.

4 Chakrabarti A, Kumar P, Padhye AA, et al. Primary cutaneous zygomycosis due to *Saksenaea vasiformis* and *Apophysomyces elegans*. Clinical Infectious Diseases 1997; 24: 580–3.

5 Weinberg WG, Wade BH, Cierny III, G, Stacy D, Rinaldi MG. Invasive infection due to *Apophysomyces elegans* in immunocompetent hosts. Clinical Infectious Diseases 1993; 17: 881–4.

6 Radner AB, Witt MD, Edwards Jr, JE. Acute invasive rhinocerebral zygomycosis in an otherwise healthy patient: case report and review. Clinical Infectious Diseases 1995; 20: 163–6.

7 Okhuysen PC, Rex JH, Kapusta M, Fife C. Successful treatment of extensive posttraumatic soft-tissue and renal infections due to *Apophysomyces elegans*. Clinical Infectious Diseases 1994; 19: 329–31.

8 Chakrabarti A, Shivaprakash MR, Curfs-Breuker I, Baghela A, Klaassen CH, Meis JF. *Apophysomyces elegans*: epidemiology, amplified fragment length polymorphism typing, and in vitro antifungal susceptibility pattern. Journal of Clinical Microbiology 2010; 48(12): 4580–5.

CHAPTER 9

Endocarditis with Unusual Organisms or Characteristics

Case 9.1   Man with dyspnea, cough, fever, and weight loss

In February, 2003, a 61-year-old Caucasian male was admitted through the emergency room (ER) with dyspnea, non-productive cough, and vague chest pain. One month earlier, he had treated himself for "cold symptoms" with over-the-counter cold medicines, and the runny nose, sneezing, and dry cough resolved.

Five to 6 days prior to admission (PTA), the non-productive cough returned, accompanied by drenching sweats and low-grade temperature. He was admitted for suspected pneumonia and to rule out myocardial infarction (MI).

The review of systems was positive for exertional dyspnea, non-productive cough, drenching sweats, low-grade fever (temperature below 101 °F) and significant weight loss of 25 pounds in 4 weeks, in spite of a good appetite. He had no headaches or generalized body aches, sore throat, diarrhea, rash or itch.

His past medical history was significant for coronary artery disease (CAD), with previous angioplasty 7 years earlier, hyperlipidemia, and hypertension. He had surgery (vagotomy) 20 years earlier.

Family history was significant for heart disease, hypertension, and stroke. A former air conditioning maintenance and repair man, the patient had retired following his heart problems several years earlier. He smoked half a pack of cigarettes for 20 years, and was not a drinker. He had no known drug allergies. Home medications included fluvastatin, diltiazem, metoprolol, and fosinopril.

Five days after admission, infectious diseases (ID) consult was requested because of positive blood cultures for *Bacillus* species.

On examination, he was alert and fully oriented, and in no acute distress. He was thin, and mild to moderately ill looking. Vital signs were as follows: blood pressure (BP) 140/70, respirations 20/min, heart rate (HR) 77, maximum temperature 100.5 °F (97 °F at time of exam); height 5′6″, weight 129 pounds.

Cases in Clinical Infectious Disease Practice: Obtaining a Good History from the Patient Remains the Cornerstone of an Accurate Clinical Diagnosis: Lessons Learned in Many Years of Clinical Practice,
First Edition. Okechukwu Ekenna.
© 2016 John Wiley & Sons, Inc. Published 2016 by John Wiley & Sons, Inc.

Head and neck exam showed upper prosthetic dentures and lower teeth in poor condition. Fundoscopy was not done. There was no jaundice, conjunctivitis, or adenopathy. The heart exam showed no murmurs, while the lung exam showed slightly increased expiration period, but no wheeze, crepitations, or consolidation. The abdominal, upper, and lower extremities, genitalia, and neurologic examinations showed nothing pathologic. Skin was free of signs of peripheral emboli or any rash, other than some forearm bruising from blood draws. Lymphadenopathy was not noted.

Laboratory and x-ray findings

Chest x-rays showed some hyperinflation and fibrous bullous emphysema, and minimal vascular congestion, but no consolidation.

Two sets of blood cultures (BACTEC system) on admission were reported positive about 60 hours later as gram-positive (GP) *Bacillus*. Additional blood cultures had been drawn to confirm that these were not contaminants. The complete blood count (CBC) showed a white blood cell (WBC) count of 15.9/µL, hemoglobin and hematocrit (H/H) 10.6 and 31.6, respectively, and platelets 415/µL. Differential count showed 82% granulocytes, 12% lymphocytes, and 5.2% monocytes. Chemistries showed a sodium value of 144, potassium 4.3, glucose 159 mg/dL, and blood urea nitrogen (BUN)/creatinine of 11 and 1.0, respectively. Erythrocyte sedimentation rate (ESR) was elevated at 72 mm/hour.

Cardiac enzymes were found to be elevated 1 day after admission, with creatine kinase myocardial band (CK-MB) and troponin I 19.6 ng/mL and 16.0 µg/L, respectively (normal range <5 and <0.42, respectively). Electrocardiogram (EKG) showed lateral T-wave inversion. Basic natriuretic protein (BNP) was elevated at 1160 pg/mL (normal range 5–100), suggestive of congestive heart failure. Nasal swab for influenza antigens was negative and sputum culture was negative for pathogens.

Hospital course and complications

The cardiologist wanted to do a cardiac catheterization (because of non-Q-wave MI), but was concerned about the reported positive blood cultures. Were these real or contaminants? The ID consultant therefore went to the microbiology lab to personally review the plates and gram stains of the positive blood cultures. The subcultured organisms on the agar plates showed adherence to the plates, and on gram stain appeared as tiny clumped GP bacilli, typical for diphtheroids or coryneform bacteria. The direct smear from the blood culture bottles showed longer chains of slightly larger gram-positive *Bacillus* (it is quite usual to expect some differences in organism morphology from broth and solid media cultures).

Since the taxonomy of non-spore-forming gram-positive rods (anaerobic or microaerophilic) is quite complex, we opted to send our specimen off to a reference laboratory for identification and susceptibility testing.

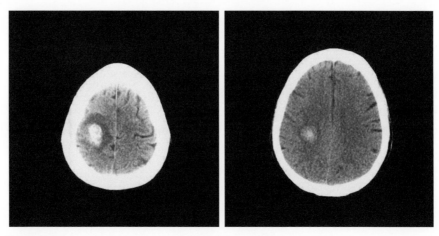

Figure 9.1a CT scan of the brain, with and without contrast, done 6 days after admission showing right superior cerebral (parietal lobe) hemorrhage.

One day after the ID consult, the patient developed left-sided weakness. Computed tomography (CT) scan of the brain confirmed hemorrhage of the right superior cerebral (parietal lobe) hemisphere (Fig. 9.1a).

Four days after the ID consult, additional studies and procedures confirmed more complications: CT of the abdomen showed lesions in the spleen and left kidney.

A CT-guided, percutaneous biopsy by radiology of the left kidney confirmed hemorrhagic necrosis, consistent with infarct.

Biopsy/aspirate of the spleen suggested an abscess: the gram stain showed numerous WBCs, rare GP bacilli, and GP cocci in chains, possibly diphtheroids. Culture of the splenic aspirate was negative.

Transthoracic echocardiogram showed no vegetations, but thickened aortic valve. The transesophageal echo (TEE) showed bicuspid aortic valve, and confirmed a 2 mm vegetation at the aortic outflow tract, and moderate aortic regurgitation.

• *How do you interpret the above findings?*

Our interpretation of the imaging and biopsy findings was that the brain, left kidney, and spleen findings reflected results of **septic emboli** from the aortic valve.

A final report of the organism identification was available from the reference laboratory about 14 days after admission. Treatment with antibiotics was continued.

Fifteen days after admission (10 days after ID consult), the patient was stabilized and sent to the hospital subacute care center for rehab for the cerebrovascular accident (CVA) and to continue antibiotic (ampicillin) therapy. Additional complications while in the rehab center included fever up to 104 °F, with negative follow-up blood cultures, relatively stable or normal labs, x-rays and general physical exam, except for a few scattered nodular, blanching exanthemata on the body torso. Vancomycin and ceftriaxone were used to replace the ampicillin, and the temperature came down to normal in 24 hours.

• *How do you interpret this new fever?*

Our interpretation is that this was most likely **drug-induced fever** due to beta-lactam drugs (ampicillin initially but also later ceftriaxone, as the fever returned at a lower grade several days later).

One month after admission, cardiac catheterization was done. It showed severe aortic regurgitation and moderate stenosis with 39 mm gradient. There was total occlusion of the right coronary artery, and other mild-to-moderate CAD findings. The renal arteries were patent, and there was normal right heart hemodynamics.

Four days after the cardiac catheterization, extraction of the remaining teeth in the lower jaw (carious teeth) was carried out.

Two weeks later, nearly 6 weeks after initial admission, surgery was done for aortic valve replacement (Edwards bovine pericardial bioprosthetic heart valve), triple-vessel coronary artery bypass graft (CABG), and left upper lobe wedge resection of lung for bullae.

The gram stain of the diseased valve was negative, and culture was also negative.

The histopathology of the calcified aortic valve tissue showed multiple diphtheroid, corynebacterial, and gram-positive coccoid organisms in short chains and singles.

Two weeks after surgery, the patient was discharged to outpatient follow-up treatment. Because of the beta-lactam allergy (drug-induced fever), his treatment was completed with clindamycin for an additional 4 months after surgery (a total duration of 6 months of antibiotic therapy).

Postscript

The patient did well, and was fully recovered. He did have an atrioventricular (AV) dual-chamber pacemaker placed for heart block 9 days after the cardiac surgery.

He was seen in my office 5 years later in 2008, for an unrelated problem, and looked healthier than he appeared to be 5 years earlier.

• *What is your final diagnosis?*

The final diagnosis was *Actinomyces israelii* aortic valve endocarditis with septic embolic complications.

Case discussion

Actinomyces spp infections are typically limited to the cervicofacial, pulmonary/thoracic, or abdominal/pelvic regions. Very rarely are these organisms associated with endocarditis. Less than 16 cases had been reported in the literature by 2002 [1,2]. These rare cases of *Actinomyces* endocarditis have a propensity for systemic embolization. Almost all have involved native valves [1].

The patient presented here had confirmed aortic valve endocarditis documented by positive blood culture for *A. israelii*, and vegetations seen on TEE. He had multiple septic embolic complications that included right parietal hemorrhage, left renal infarct, and splenic abscess.

Although the diseased valve was culture negative, the histopathology confirmed the presence of gram-positive organisms consistent with actinomycosis. The negative valve culture is consistent with the effective antibiotic therapy received in the previous weeks before valve surgery.

This patient had additional complications during therapy, including heart block that required placement of an AV dual-chamber (DDD) pacemaker.

He also developed a fever during appropriate antimicrobial therapy with ampicillin that was most likely drug-induced fever related to the beta-lactam antibiotic. The timing of the fever (2 weeks into therapy) fits with this. For that reason, he was eventually discharged home to complete the remaining 4 months of a total of 6 months of antibiotic therapy, with oral clindamycin.

In this patient, the presumptive portal of entry of the *Actinomyces* was the poor lower jaw dentition, although we have no definitive proof of this. The remaining lower set of teeth was removed before the valve replacement surgery.

The patient's treatment was very successful. He made a complete recovery. He had no obvious residual of the CVA when he walked into my office for an unrelated problem in 2008, 5 years after completion of the endocarditis treatment.

General diagnostic and treatment comments

Even today, the diagnosis of actinomycosis remains a difficult challenge. Most strains of *Actinomyces* are microaerophilic or facultative (except *A. meyeri*) anaerobes. However, strict anaerobic processing and conditions should be utilized for primary isolation. Although macroscopic grains or "sulfa granules" are a very helpful diagnostic clue, this is not seen in more than 25% of cerebral confirmed cases of actinomycosis [3]. Their absence should not be used to rule out actinomycosis.

The full identification and speciation of non-spore-forming GP rods (anaerobic or microaerophilic) is complex, and beyond the competence of most community laboratories [4]. Sterile source isolates of such agents should be sent to a reference lab for identification and susceptibility testing. Most good labs utilize 16S ribosomal RNA gene sequencing as an additional aid in speciation. This method has led to the reclassification of many of the "classic" *Actinomyces* species [5].

There were no established ranges for the Kirby–Bauer (KB) disk testing at the time of patient evaluation in 2003. However, the susceptibility of *Actinomyces* spp to penicillin, ampicillin, tetracycline/doxycycline, minocycline, and clindamycin is predictable [5]. Other antimicrobial agents (ceftriaxone) have been used successfully, anecdotally, or have documented good in vitro activity, and are likely to be effective against the *Actinomyces* (linezolid, vancomycin, moxifloxacin). The organism is usually very sensitive to many antimicrobial agents, and a single dose of an effective agent can hinder the isolation of *Actinomyces*. It is important, therefore, to avoid any antimicrobial treatment before obtaining specimens for culture, in order to enhance the chances of recovery of these organisms.

Most experts recommend prolonged treatment (6–12 months) for serious infections with actinomycosis.

Lessons learned from this case

- *Actinomyces* endocarditis is a very rare but serious disease fraught with complications.
- Transesophageal echocardiogram (TEE) may be required to confirm heart valve vegetations, as in this case.
- The presence or occurrence of multiple systemic septic emboli did not reflect a failure of therapy, but was part of the typical clinical manifestations of this subacute disease.
- Definitive identification of this organism may require expert help (reference laboratory), because of the difficult taxonomy of these gram-positive non-spore-forming microaerophilic/anaerobic organisms.
- The susceptibility of this group of organisms (*Actinomyces*) is predictable. Alternatives to penicillin/ampicillin are available (tetracycline/doxycycline, minocycline, clindamycin).
- Drug-induced fever can occur (as happened in this case), and should not be construed as failure of therapy. In this patient, an alternative drug (clindamycin) was used to complete the treatment.
- Prolonged treatment with antimicrobial agents is required for complete cure of this infection.
- The initial report of a *Bacillus* or diphtheroid from a blood culture should not be dismissed as a contaminant without full assessment of the clinical, laboratory, and epidemiological circumstances.

References

1 Mardis JS, Many WJ. Endocarditis fue to *Actinomyces viscosus*. Southern Medical Journal 2001; 94(2): 240–3.

2 Westling K, Lidman C, Thalme A. Triscupid valve endocarditis caused by a new species of Actinomyces: *Actinomyces funkei*. Scandinavian Journal of Infectious Diseases 2002; 34(3): 206–7.

3 Winkling M, Deinsberger W, Schindler C, Joedicke A, Boeker DK. Cerebral manifestation of an actinomycosis infection. A case report. Journal of Neurosurgical Sciences 1996; 40(2): 145–8.

4 Moncla BJ, Hiller SL. Peptostreptococcus, Propionibacterium, Lactobacillus, Actinomyces, and other non-spore-forming anaerobic gram-positive bacteria. In: Murray PR, Baron EJ, Jorgensen JH, Pfaller MA, Yolken RH (eds) *Manual of Clinical Microbiology*, 8th edn. Washington, DC: ASM Press, 2003, pp. 857–79.

5 Russo TA. Agents of actinomycosis. In: Mandell GL, Bennett JE, Dolin R (eds) *Principles and Practice of Infectious Diseases*, 7th edn. Philadelphia: Churchill Livingstone, 2010, pp. 3209–19.

Further reading

Apothéloz C, Regamey C. Disseminated infection due to *Actinomyces meyeri*: case report and review. Clinical Infectious Diseases 1996; 22: 621–5.

Case 9.2   Confusing staining characteristics and false identification of a pathogen by automated system in a seriously ill patient

A 59-year-old Caucasian male was admitted in late June 1999 with fever, malaise, and weight loss.

Over the preceding 4 months, he had complained of generalized malaise and fatigue, and non-specific aches. Later, he developed muscle and joint pains involving the knees and shoulders. He was referred to a rheumatologist and received non-steroidal anti-inflammatory drug (NSAID) treatment. Progressive anemia was also noted by the primary care physician. Three to 4 weeks prior to admission, low-grade fever, in association with sweating, especially following NSAID and acetaminophen therapy, was observed by the patient. Although he continued to go to work, he felt generally run down and tired. One week before admission, his wife recorded a temperature of 102.4 °F. At this time, inpatient admission was sought for more complete investigation.

The past medical history was significant for hypertension and prostate cancer (T2, N0, M0), treated with surgery in 1997, and no additional therapy, but monitoring with prostate-specific antigen (PSA). He had renal stones many years previously, vertigo since youth (inner ear problem), and anemia for which he had been on ferrous sulfate and multivitamin supplements. In 1994 and 1997, echocardiogram confirmed mitral regurgitation murmur.

He had a history of type 1 allergy to penicillin (throat, facial, and body swelling): anaphylaxis.

He was married to a healthy wife, and they had a 20-year-old daughter. They lived in a semi-rural area of South Mississippi. An electrician, he did not smoke or drink alcohol. At home, they had an outside cat, 10 years old, and no other animals or birds. There had been no recent dental work done on him.

A weight loss of 30–36 pounds in the preceding 6 months was noted by his family physician. Other symptoms in the review of systems are as outlined above.

He was on the following medications prior to admission: fosinopril, ferrous sulfate supplements, some NSAID, and acetaminophen.

Two days after admission, infectious diseases consultation was placed for fever, chills, weight loss, and possible vegetation on transthoracic echocardiogram.

On examination, he was alert, oriented, co-operative, pleasant, but tired and worried. His vital signs were as follows: BP 102/56, respiratory rate (RR) 17, HR 86, temperature 100.3 °F (T_{max} 102.7 °F on 6/22/99).

Head and neck examination showed no conjunctivitis or jaundice. On fundoscopy, no vascular lesions were noted. Oral exam showed caries of the molar teeth, while no adenopathy in the neck was noted. The heart exam showed systolic ejection murmur in the aortic and pulmonary areas but pansystolic murmur, grade 3–4/6, radiating to the axilla and consistent with mitral regurgitation. The lungs were clear, and the abdominal, extremities, genitalia, skin, neurologic, and lymph node exams were negative. There were no splinters of the nails or skin rash noted.

- *What differential diagnoses would you consider at this time, even before any lab results?*

Laboratory and other investigations

Prior to the hospital admission on 6/21/99, the patient had already had some work-up done. Blood cultures were done on admission, but the results were still negative at 2 days on 6/23/99 when the infectious diseases physician was consulted.

Bone scan showed old trauma of the left ankle; EKG and chest x-ray were normal. Abdominal ultrasound showed non-specific changes in the gallbladder area. Iron studies confirmed low saturation of 7%, low concentration of 22%, and normal total iron binding capacity (TIBC) and ferritin levels. B12 and folate levels were normal, and reticulocyte count was 2.0, slightly elevated. The CBC showed a WBC count of 7.3, H/H 11.7/35.1, mean corpuscular volume (MCV) 72.1, platelet count 113,000. The differential count showed 77% polymorphs, 12% lymphocytes, and 10% monocytes. The ESR was 48 mm/hour. Urinalysis was normal, showing no red cells or casts; PSA was normal, and carcinoembryonic antigen (CEA) was normal. BUN/creatinine were 16/0.9, respectively; alkaline phosphatase was normal (108), but lactate dehydrogenase (LDH) was elevated at 835 (upper limit 618). Sodium was 139, potassium 4.2, glucose 93 mg/dL. Antinuclear antibody (ANA) test was negative.

Echocardiogram done on 6/22/99 was suspicious for a vegetation in the anterior mitral leaflet. Three sets of blood cultures were drawn on 6/21/99, and another five sets on 6/22/99. All eight sets were negative on 6/23/99.

Hospital course

After the blood cultures were drawn, the patient was started on intravenous vancomycin, especially because of the history of serious penicillin allergy. Later during the night of 6/23/99, the blood culture came up positive, and a subculture was plated onto several solid media for identification. The next day, 6/24/99, there was no growth on any media except the chocolate blood agar (CBA). The gram stain from the CBA showed pleomorphic gram-negative bacilli.

• *What is the likely organism identified on this gram stain?*

The patient was doing reasonably well clinically on vancomycin therapy but we now have gram-negative rods reported on the stain.

• *Would you change therapy at this time?*

We opted to wait another 24 hours to see if there would be growth the next day on other solid culture media, including MacConkey agar, a differential and selective medium for gram-negative organisms. We discussed possible differentials with the laboratory staff. HACEK (*Haemophilus, Actinobacillus, Cardiobacterium, Eikenella*, and *Kingella*) organisms do not grow this fast (overnight), and were considered unlikely. While final identification was pending in the automated ID system, we asked for KB susceptibility tests for both gram-positive and gram-negative organisms to be done.

Vancomycin was continued, and gentamicin added to the regimen for the patient. TEE was done, and confirmed multiple vegetations involving the following valves: multiple small size lesions on the left and non-coronary cusp of the aorta and on both the mitral valve leaflets, with the largest 1.3 × 0.45 cm, and two small vegetations on the posterior mitral annulus.

The next day, 6/25/99, there was still no growth on any other solid media, except the CBA, but the KB disk sensitivity report (done on CBA) was available, and showed susceptibility to the following agents: ampicillin, penicillin, ceftriaxone, vancomycin, chloramphenicol, clindamycin, levofloxacin, ciprofloxacin, tetracycline, gentamicin, and rifampin. It was resistant to trimethoprim/sulfamethoxazole.

The organism was identified by the "Miroscan" automated system as either *Neisseria* or *Vibrio*. That organism ID report made no sense and so could not be accepted.

At this stage, the ID consultant asked to personally review the original BC broth growth on gram stain. A fresh gram stain from one of the positive BC bottles was prepared.

A presumptive diagnosis was made, and a streaking procedure on a fresh sterile blood agar (BA) plate was requested, using the growth from the BC bottle directly, as well as the growth on the CBA. The plates were to be viewed the next day in the lab.

The presumptive diagnosis was confirmed the next day on review of the streaked BA plates.

- *Can you guess what was seen on the gram stain from the BC bottle, and what identification was confirmed by the streak on the solid BA media?*
- *What organism was used to do the streaking on the BA plate?*

The patient did well and was discharged to outpatient follow-up 12 days after admission on 7/3/99, on a combination of vancomycin and gentamicin, after a central line was placed on 7/2/99.

Figure 9.2a shows a graphic display of the temperature recordings from 6/21 to 6/30/99. The temperature became normal on 6/25/99, 4 days after admission.

Case discussion and follow-up

This patient had subacute bacterial endocarditis (SBE). The diagnosis was confirmed by positive blood cultures (8 of 8), and by echocardiograms (TTE and TEE) showing multiple vegetations. He had pre-existing mitral valve disease in the form of mitral regurgitation confirmed by previous echo in 1994 and 1997 before he became ill. He had been ill for nearly 4 months prior to admission with malaise, weight loss, and later fever, etc. Although he had no recent dental work done, he had poor dentition (caries) as noted in the physical examination. He fulfilled the required major and minor criteria for the diagnosis of infective endocarditis [1].

Because he had known allergy to penicillin, he was started on vancomycin after BCs were drawn. The positive BC was subcultured, as is normal practice, to several solid media (selective and differential). It showed no growth on any of the media except CBA. The gram stain from the CBA showed pleomorphic gram-negative bacilli. It was only when the original BC bottle growth was

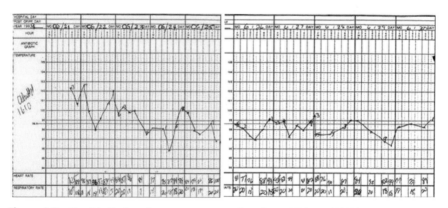

Figure 9.2a Graphic display of temperature recordings from 6/21 to 6/30/99.

reviewed did it become clear that there was a discrepancy in the organism identification. The gram stain from the positive BC bottle (broth medium) showed gram-positive cocci in chains. This suggested that we were likely dealing with a nutritionally deficient (variant) *Streptococcus* (NVS). That was the presumptive diagnosis that was made, hence the confidence in continuing vancomycin. The KB susceptibility tests reconfirmed and reassured us that this was indeed a gram-positive organism. To prove that we were dealing here with a NVS, the laboratory was asked to streak a *Staphylococcus* organism across a BA plate inoculated with the NVS. The next day there were tiny "satellite" organisms around the *Staphylococcus aureus* streak.

An example of the type of observed "satellite" growth around a *Staphylococcus aureus* streak is shown in Fig. 9.2b [1].

- *Why did the organism grow in the blood culture bottle (broth medium) and not on the solid BA media?*

The BC medium has growth factors needed by this organism. The system in use at the hospital was BACTEC, which contained pyridoxal (vitamin B6), a growth factor needed by this NVS. The growth was possible because the CBA contains partially broken down hemoglobin, providing some of the growth factor needed by the NVS, but not the same quality as the liquid media, hence the gram-negative nature of this entity previously classified as *Streptococcus*. The NVS have now been reclassified under two new genera: *Abiotrophia* and *Granulicatella* [2].

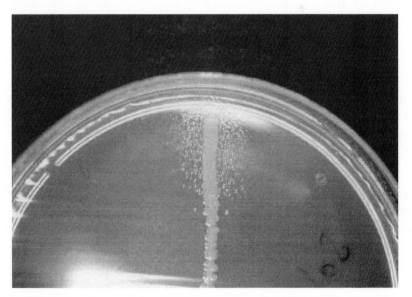

Figure 9.2b *Abiotrophia* spp (formerly known as nutritionally deficient streptococci) showing satellite growth with *Staphylococcus aureus*. Reproduced from Brouqui and Raoult [1], with permission from the American Society for Microbiology.

A brief review of the characteristics of the NVS will be given after the additional follow-up report on this patient is reviewed.

In- and outpatient follow-up and complications of therapy

The patient was hospitalized for 12 days, from 6/21 to 7/3/99. He returned to the hospital for 2 days from 7/11 to 7/13/99 because of fever (101.8°) and lower abdominal pain. Blood cultures were negative, including from the central line. Stool studies were negative for *C. difficile* toxin, enteric pathogens, ova, cysts, and parasites. A new central line was placed and the old one removed. There were no further admissions to the hospital.

He was seen several times as an outpatient, and labs were monitored closely. Gentamicin was discontinued on 7/16/99, after 3 weeks, because of increasing creatinine levels (BUN/creatinine were 40/2.8 on 7/15/99). Rifampin was started on 7/21/99 and stopped after 2 days because of nausea, vomiting, and "drunken feeling." A third central line (PICC) was placed on 7/23/99 for the rest of the vancomycin that ended after 5.5 weeks total. Because of increasing dizziness, ear, nose, and throat (ENT) referral was done. Brain/skull magnetic resonance imaging (MRI) was negative. High-frequency sensorineural deficit – hearing loss – was confirmed. Note that this patient had a long history of vertigo since his youth.

In September, 1999, at 6 weeks post completion of therapy, he had no further complaints. He had gained 25 pounds (weight 219 pount). His examination was normal; he still had his old MR murmur, now grade 3/6. The labs showed a WBC count of 5.9, H/H 13.5/40.6, MCV 82.1; BUN/creatinine levels were normal at 16/1.3, respectively. The ESR was also normal at 8 mm/hour. Follow-up BCs done a few weeks earlier were negative. He was discharged from the clinic.

• *Final diagnosis: Nutritionally variant (deficient) Streptococcus bacterial endocarditis.*

Postscript

The patient had his mitral valve replaced 1 month later, on 10/4/99. I spoke with him on the phone on 6/20/11, nearly 12 years after his endocarditis and surgery. He was alive and doing well. He was 72 years old in October, 2011.

Summary of characteristics of nutritionally variant (deficient) *Streptococcus* (NVS)

The NVS were first described in 1961 by Frankel and Hirsch as fastidious streptococci that only grew as satellite colonies around other bacteria [3]. Originally isolated from cases of bacterial endocarditis and otitis media, the authors thought these organisms were mutant forms of *S. mitis*. Many reports have since been published describing these organisms as causes of human disease, especially endocarditis [4,5] and brain abscess [6]. Some of these publications also reflect the flux in nomenclature or methods of identification [7–11].

Over the last two decades, however, based on DNA-DNA hybridization studies and 16S ribosomal RNA sequence analysis, these organisms have now been placed into two genera (*Abiotrophia* and *Granulicatella*), comprising four species that have been identified from human specimens [2]. These four species are *Abiotrophia defectiva*, *Granulicatella adiacens*, *G. elegans*, and *G. para-adiacens*.

All NVS species require pyridoxal or cysteine for growth. There appears to be enough pyridoxal in human blood to support the growth of NVS in most blood culture systems, as was the case in our BACTEC blood culture method used to grow this organism. For subculture, however, solid media must be supplemented with pyridoxal or cysteine to sustain growth. Only the CBA showed some growth of our patient's BC isolate, while other media did not support growth of the NVS. Even the growth on the CBA was defective, showing pleomorphic gram-negative bacilli instead of gram-positive cocci. When the subculture plate was cross-streaked with *S. aureus* to provide the missing factors, growth was permitted as satellite colonies around the streak. This was the procedure performed with our NVS on 6/25/99, and the basis for the presumptive diagnosis [11].

The NVS produce similar diseases to other viridans streptococci, including infective endocarditis. They cause about 5% of cases of bacterial endocarditis [2]. Infective endocarditis caused by NVS is virtually always indolent in onset and associated with prior heart disease (as was the case in our patient). Endocarditis caused by NVS also carries greater morbidity and mortality than that caused by other streptococci [2]. The NVS show tolerance to penicillin, but all strains are susceptible to vancomycin. Our patient was allergic to penicillin, and so received vancomycin initially and for the rest of the prolonged 5 plus weeks of therapy.

The NVS, like viridans streptococci, are found as normal flora in the upper respiratory, urogenital, and gastrointestinal tracts of humans. They are usually alpha-hemolytic or non-hemolytic, and show pleomorphic or variable morphology on gram stain in media lacking growth factors [11].This is the explanation of why the initial solid media gram stain was confusing. The stain done subsequently from the positive blood culture (liquid, broth) medium showed gram-positive cocci in pairs and chains.

Lessons learned from this case

- Nutritionally variant streptococci (NVS) are responsible for up to 5% of culture-negative endocarditis.
- Infective endocarditis caused by NVS is typically almost always indolent in onset, and associated with prior valvular heart disease (as in our patient).
- NVS cause a disproportionately greater amount of complications than regular viridans streptococci that include embolization, a higher mortality, and also relapse.
- They are difficult to diagnose because of their special growth requirements (e.g. L-cysteine or pyrodoxine, and other growth factors).

- The staining and other characteristics of these organisms vary depending on the available growth factors and the type of medium they are grown in.
- Automated systems are not very useful in identifying these organisms, because of their special growth factor requirements, and because they are uncommon.
- The nomenclature of NVS has been in flux, as new genetic methods of identification are introduced.
- The NVS remain typically susceptible to vancomycin, even as some may show tolerance to penicillin.

References

1 Brouqui P, Raoult D. Endocarditis due to rare and fastidious bacteria. Clinical Microbiology Reviews 2001; 14(1): 177–207. Available at: www.ncbi.nlm.nih.gov/pmc/articles/PMC88969/, accessed March 2, 2016.

2 Sinner SW, Tunkel AR. Nutritionally variant (deficient) Streptococci (*Abiotrophia* and *Granulicatella*). In: Mandell GL, Bennett JE, Dolin R (eds) *Principles and Practice of Infectious Diseases*, 7th edn. Philadelphia: Churchill Livingstone, 2010, p. 2673.

3 Frenkel A, Hirsch W. Spontaneous development of L forms of streptococci requiring secretions of other bacteria or sulphydryl compounds for normal growth. Nature 1961; 191: 728–30.

4 Bouvet A. Human endocarditis due to nutritionally variant streptococci. European Heart Journal 1995; 16(suppl. B): 24–7.

5 Roggenkamp A, Abele-Horn M, Trebesius KH, Tretter U, Autenrieth IB, Heesemann J. *Abiotropia elegans* sp. Nov., a possible pathogen in patients with culture negative endocarditis. Journal of Clinical Microbiology 1998; 36(1): 100–4.

6 Biermann C, Fries G, Jenichen P, Bhakdi S, Husman M. Isolation of *Abiotrophia adiacens* from a brain abscess which developed in a patient after neurosurgery. Journal of Clinical Microbiology 1999; 37(3): 769–71.

7 Ohara-Nemoto Y, Tajika S, Sasaki M, Kaneko M. Identification of *Abiotrophia adiacens* and *Abiotrophia defectiva* by 16S rRNA gene PCR and restriction fragment length polymorphism analysis. Journal of Clinical Microbiology 1997; 35(10): 2458–63.

8 Roggenkamp A, Leitriiz L, Baus K, Falsen E, Heesemann J. PCR for detection and identification of *Abiotrophia* spp. Journal of Clinical Microbiology 1998; 36(10): 2844–6.

9 Beighton D, Homer KA, Bouvet A, Storey AR. Analysis of enzymatic activities for differentiation of two species of nutritionally variant streptococci, *Streptococcus defectivus* and *Streptococcus adjacens*. Journal of Clinical Microbiology 1995; 33(6): 1584–7.

10 Kawamura Y, Hou XG, Sultana F, Liu S, Yamamoto H, Ezaki T. Transfer of *Streptococcus adjacens* and *Streptococcus defectivus* to *Abiotrophia* gen. nov., respectively. International Journal of Systematic Bacteriology 1995; 45(4): 798–803.

11 CDC. Streptococcus Laboratory General Methods: Pyridoxal Requirement Test (p. 27); Satellite Test (p. 30); and Table: biochemical differentiation of nutritionally variant streptococci (NVS) species (p. 56). Available at: www.cdc.gov/streplab/downloads/general-methods-sections1-2.pdf, accessed March 2, 2016.

Further reading

Durack DT, Lukes AS, Bright DK. New criteria for diagnosis of infective endocarditis: utilization of specific echocardiographic findings: Duke Endocarditis Service. American Journal of Medicine 1994; 96: 200–9.

Murray PR, Baron EJ, Jorgensen JJ, Pfaller MA, Yolken RH. *Manual of Clinical Microbiology*, 8th edn. Washington, DC: ASM Press, 2003.

CHAPTER 10

Severe Systemic Fungal and Other Infections in AIDS Patients

Case 10.1 HIV patient with progressive shortness of breath

A 37-year-old Caucasian male was admitted in January, 2007 with a 4-week history of progressive shortness of breath (SOB). His initial symptoms were described as flu-like, with vague sore throat that failed to respond to two courses of amoxicillin and amoxicillin/clavulanate therapy. He continued to have cough productive of whitish "phlegm." Three days before admission, he had come to the emergency room (ER) for SOB and was prescribed trimethoprim/sulfamethoxazole (TMP-SMX). That treatment did not help, so he returned to the ER and was admitted.

He was initially admitted to the regular floor, but subsequently transferred to the medical intensive care unit (MICU) the same day because of hypoxemia (pO$_2$ in the 40s and O$_2$ sat of 84%).

The next day, infectious diseases consult was placed because of abnormal chest x-ray and progressive SOB, as bronchoscopy was being done on the patient. The bronchoscopy findings were described as showing diffuse mucosal inflammation and thin watery secretions, with no endobronchial lesions or purulence. No biopsy was done, but bronchoalveolar lavage (BAL) for multiple stains and cultures was carried out.

The review of systems was significant for the following: progressive SOB for 4 weeks; flu-like symptoms and sore throat for several weeks; non-productive, non-purulent cough; nose bleed only 2 days prior to admission from coughing and sneezing; poor appetite for weeks; weight loss of 17 pounds in 1 month; thrush for 3 weeks (improved on nystatin and TMP-SMX); and fever for 3–4 weeks, with fatigue.

His past medical history was significant for HIV infection diagnosed in 1995 (12 years earlier). He received antiretroviral therapy for several years (1995–2000), and stopped for reasons that were unclear. He apparently

Cases in Clinical Infectious Disease Practice: Obtaining a Good History from the Patient Remains the Cornerstone of an Accurate Clinical Diagnosis: Lessons Learned in Many Years of Clinical Practice, First Edition. Okechukwu Ekenna.
© 2016 John Wiley & Sons, Inc. Published 2016 by John Wiley & Sons, Inc.

retested himself (oral mucosal swab that was then sent by mail to a remote laboratory), and was found to be negative.

The social and family history was significant for the following: he was homosexual, single, and had no children. His partner had died in 1995 when he was diagnosed with HIV infection, although he was not ill at the time. He had lived in Michigan at the time, and moved to the Gulf Coast 3 years earlier. There was no significant family history other than some cancer in the mother. He was not a smoker and did not drink alcohol. He had no known drug allergies.

He was taking only trimethoprim/sulfamethoxazole, prescribed 3 days before admission.

On examination, he was in mild-to-moderate distress, had a 100% rebreather, with the head of the bed elevated. His vital signs were as follows: blood presssure (BP) 121/65, respiratory rate (RR) 35/min, heart rate (HR) 118/min, temperature 99.2°F (102.6°F the day before); height 5'9", weight 183 pounds (83 kg).

The head and neck exam showed no jaundice, conjunctivitis or thrush; the heart and lung exams were unremarkable. Abdomen was protuberant, full, and otherwise normal. Upper and lower extremities were unremarkable; lymph nodes were not palpably enlarged in the neck, axilla, or groin. External genitals showed a normal circumcised male phallus; the skin showed no rash, but there were tattoos on the torso as well as upper and lower extremities. Neurologic exam was normal. Rectal exam was not done.

Laboratory and x-ray findings showed the following: complete blood count (CBC) showed a white blood cell (WBC) count of 6800/µL, hemoglobin and hematocrit (H/H) 13.6/39.8, respectively, mean corpuscular volume (MCV) 85.4, and platelets 104,000. Differentials showed 74% polymorphs, 14% band forms, 9% lymphocytes, and 3% monocytes. Chemistries showed a sodium of 136, potassium 3.9, glucose 92, blood urea nitrogen (BUN)/creatinine 11/1.0, respectively; alkaline phosphatase was 268, total bilirubin was 1.1, albumin 2.5, and lactate dehydrogenase (LDH) was 623. Blood cultures drawn on three separate occasions on 1/5, 1/8/and 1/14/07 were negative. Mycoplasma and Monospot tests were negative. Gallbladder ultrasound was negative for stones or disease.

Chest x-ray showed interstitial and nodular lung disease without consolidation (Fig. 10.1a).

Hospital course and additional tests

One day after admission, the day of the initial bronchoscopy, additional labs showed the following: WBC 5.3, H/H 10.9/32.2, respectively, platelets down to 59,000; differential count showed 37% polymorphs, 50% bands, 5% lymphocytes, 3% monocytes, and 4% myelocytes. Sodium was down to 125 mEq/L, potassium 3.9, glucose 154 mg/dL, BUN/creatinine 14/1.2, respectively; elevated

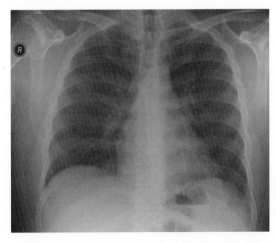

Figure 10.1a Chest x-ray taken on 1/5/07 during the initial ER visit, 3 days before admission.

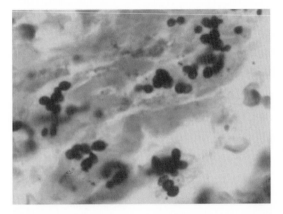

Figure 10.1b GMS stain of RLL transbronchial biopsy done on 1/11/07. Magnification ×40: yeast forms, average size 2.5 microns. (*See insert for color information*)

serum glutamic oxaloacetic transaminase (SGOT) and serum glutamic pyruvic transaminase (SGPT) (475/197, respectively), alkaline phosphatase 231, total bilirubin 2.9, and albumin 1.5 mg/dL. Prothrombin time (PT) was elevated at 18.9 seconds, partial thromboplastin time (PTT) was 46 seconds. CD4 count was very low, 9/μL, and HIV-1 viral load was 97,096 copies/mL.

Two days later, a repeat bronchoscopy was done, this time with transbronchial biopsy. Only one of the multiple stains (a biopsy from the right lower lung) was positive for intracellular yeast form organisms on histopathology (Fig. 10.1b).

At discharge on 1/24/07 (16 days after admission), the patient was clinically improved; all blood and lung tissue biopsy cultures were still negative. The epidemiologic history was again reviewed, additional serology ordered, and patient was discharged on a presumptive diagnosis on "specific" antifungal therapy, to outpatient follow-up. TMP-SMX therapy was also continued.

Second admission 5 weeks later: March 1–15, 2007

The patient was readmitted on 3/1/07 with fever (temperature 103.8°F), malaise, and cough with occasional brownish phlegm. Symptoms had been ongoing for 4 days, and had started 10 days after completion of prescribed medication in mid-February. He did not have money to refill his medication, and missed his scheduled outpatient follow-up appointment.

The infectious disease physician was reconsulted the next day to review the patient.

On examination, the patient was in no acute distress; vital signs showed the following: BP 127/83, RR 20, HR 115, temperature 102.1°F; weight 176.5 pounds (a loss of 7 pounds from January, 2007). The head and neck exam showed a purplish violaceous lesion in the hard palate on the left, reminiscent of Kaposi's sarcoma (Fig. 10.1c). There was no skin rash. The rest of the exam showed nothing else remarkable, and not different from 5 weeks earlier.

He was placed on multiple antimicrobial agents by the attending physician, designed to cover sepsis from a wide range of pathogens: levofloxacin, vancomycin, meropenem, caspofungin, intravenous TMP-SMX, and methylprednisolone.

Three days later, his clinical status deteriorated, with increased dyspnea and respiratory distress (pO_2 65%, pCO_2 24, O_2 sat 94.8% on 100% oxygen rebreather), increased micronodular chest x-ray changes (Fig. 10.1d), and

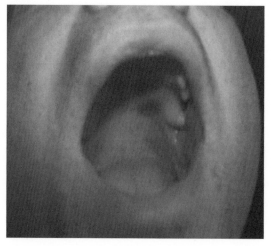

Figure 10.1c Purplish violaceous lesion in the hard palate noted on 4/5/07 (reproduced with permission).

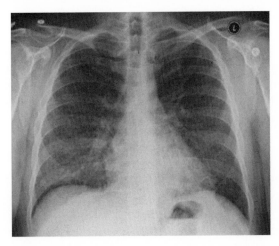

Figure 10.1d Deterioration of chest x-ray compared to January, 2007: micronodular/miliary pattern, 3/5/07.

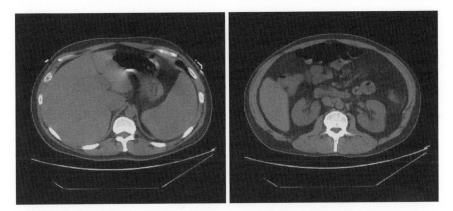

Figure 10.1e CT scan of abdomen and pelvis showing splenomegaly and enlarged lymph nodes in the pelvic region, 3/6/07.

persistent fever (temperature 102.1 °F). Enlarged pelvic lymph nodes and splenomegaly were noted on CT scan of the abdomen and pelvis (Fig. 10.1e).

He was transferred the next day to the MICU.

Major modifications in therapy

In the MICU, a peripherally inserted central venous catheter (PICC) was placed, and lipid amphotericin B was started in place of itraconazole. Steroid (methyl-prednisolone), vancomycin, meropenem, levofloxacin, and TMP-SMX were discontinued. He was continued on 100% rebreather in the MICU, and did not require intubation.

Three days later, he was transferred back to the regular floor, markedly improved. Highly active antiretroviral therapy (HAART) was started on 3/12/07, 6 days after transfer out of the MICU. The next day, his oxygen saturation on room air was 97%. On that same day (3/13/07), amphotericin B was discontinued, in place of oral itraconazole. The PICC line was removed the next day (day 8) and the patient discharged home on 3/15/07 (day 10), to continue oral therapy with HAART and itraconazole.

• *What is the likely diagnosis?*

Confirmation of diagnosis

The fungal blood culture of 1/9/07 took 5–6 weeks to be reported positive, and required a few more weeks of subculture to be confirmed and identified. The right lower lung biopsy of 1/11/07 was also confirmed positive for mold 5 weeks later, but required additional work-up, including tease mount and use of differential media. On 3/21/07, a 21-day-old mold culture was sent off to a reference laboratory for confirmation by DNA probe (Figs 10.1f, 10.1g).

Complement fixation antibodies were positive at a titer of 1:64 for *Histoplasma* mycelial phase, but not for the yeast phase antibodies (normal <1:8). The DNA probe of the blood culture specimen was confirmed positive for *Histoplasma capsulatum* within 4 days of receipt of our 21-day-old culture. The organism was sensitive to both amphotericin B and itraconazole.

This patient completed 3 months of itraconazole and stopped. He has been followed regularly in the office every 6 months, and was last seen in late 2015. He remains very healthy, has undetectable HIV-1 viral load, and adequate CD4 count. His labs and x-rays are normal, and he has been off the antifungal regimen for 8 years. The purplish violaceous hard palate lesions have since resolved. That lesion was likely due to disseminated fungal (*Histoplasma*) infection, and not to Kaposi's sarcoma.

Figure 10.1f 21-day-old mold culture on brain heart infusion (BHI) and Sabouraud dextrose agar (SDA). (*See insert for color information*)

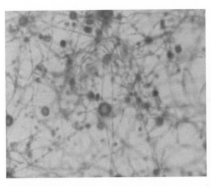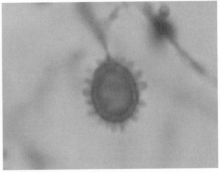

Figure 10.1g Slide/tease mount of 21-day-old culture with the larger macroconidia and the smaller microconidia evident in this mycelial phase *Histoplasma*; the macroconidium or tuberculate form with protrusions is also shown in the next picture. The macroconidia are usually 8–15 microns in size, while the microconidia are about 2–5 microns. (*See insert for color information*)

- *Final diagnosis: Disseminated histoplasmosis (DH) in an immunocompromised AIDS patient with:*
  - *positive blood and lung biopsy cultures*
  - *enlarged and bulky lymph nodes within the iliac region of the pelvis*
  - *splenomegaly*
  - *reactive small hilar mediastinal adenopathy consistent with DH*
  - *positive lung biopsy histology for yeast forms consistent with Histoplasma (H&E and GMS stains).*

Case discussion

This case illustrates many important findings that begin with an appropriate and detailed history of the present illness, the relevance of epidemiology in reaching a diagnosis, patience in arriving at a microbiologic diagnosis, as well as social issues that affected the presentation of the patient's illness.

This patient was diagnosed with HIV infection in 1995, 12 years before he presented with disseminated histoplasmosis. He had taken his HIV-specific medications for 5 years and stopped in 2000. It is not clear whether not having health insurance had anything to do with his decision to stop his HAART, but he had also apparently self-tested for HIV (in spite of a well-documented previous diagnosis) by sending off a mucosal swab to a mail-in distant laboratory that allegedly told him he was "negative." Maybe he wanted to believe that, but clearly he was not negative and was still positive. He went along for 7 years without specific HAART, and thus became severely immunosuppressed, predisposing him to opportunistic infection with *Histoplasma*. The CD4 count of 9/µL and his HIV-1 viral load of 97,096 copies/mL provide proof of his severely depressed immunity.

By the time he was seen by us, he had lived on the Gulf Coast for 3 years. Without the knowledge of his previous abode in Michigan (an endemic area for histoplasmosis), the diagnosis may not have been made early enough during the first hospital admission. Unfortunately, he stopped his itraconazole prematurely because of lack of insurance and understanding of the need for prolonged therapy. He failed to show up for his follow-up office appointment, but returned to the hospital in March when he became very ill with dyspnea and fever.

We had started him on itraconazole presumptively, based on the yeast forms on histopathology of the transbronchial biopsy in January. The culture results (lung and blood) were confirmed many weeks later. During his second admission, he received lipid amphotericin B initially, with rapid improvement of symptoms. He was then discharged home on oral itraconazole, which he completed appropriately. HAART was started, and his HIV viral load became undetectable within 3 months, and his CD4 count rose to above 200 within several months. He has remained clinically stable and off antifungal therapy for over 8 years. He should do well as long as his immune status is good.

In this patient, *Histoplasma* was most likely acquired while he lived in Michigan, and his disease was an endogenous reactivation of a latent infection. He was predisposed because of a severely compromised cell-mediated immune system. Control or improvement of the immune status is the key to keeping him free of this opportunistic infection.

Endemic areas for histoplasmosis include the Mississippi, Ohio, and Lawrence River valleys in the United States. Other areas outside the US include the Caribbean, southern Mexico, and certain parts of Central and South America, Africa, and Asia [1]. The history of where patients have lived and traveled is important in making an early diagnosis, since the culture takes up to 8 weeks to confirm. Needless to say, mortality is much less with early treatment (can be up to 50% in severe disease in AIDS patients) [1].

The differential diagnoses of the interstitial and reticulonodular infiltrates of the chest x-ray in this patient included *Pneumocystis jiroveci* (formerly PCP), tuberculosis, and viral pneumonia. However, the lung biopsy findings (intracellular yeast) and the epidemiology about where he lived in the past helped in arriving at a diagnosis before the culture results were available.

Histoplasma capsulatum, like *Coccidioides immitis* and *Blastomyces dermatitidis*, is a thermally dimorphic fungus, which means that it appears in yeast forms in tissue (or body temperature) and in mycelial (mold) forms in the environment. The culture agar used to grow this organism tries to mimic these conditions. The brain heart infusion (BHI) agar (with blood) was incubated at 37°F, while the Sabouraud dextrose agar (SDA) was incubated at 30°F (room temperature). As was seen above, the culture plates reflected more yeast in the BHI and more mold in the SDA. The lung tissue biopsy clearly showed yeast forms, as was also seen in the H&E stain of the biopsy (not shown).

Lessons learned from this case
- The importance of a good history and physical examination in making a diagnosis.

- The relevance of epidemiology in arriving at an early or presumptive diagnosis, before final laboratory confirmation of the diagnosis (knowledge of diseases found in certain geographic areas can be important in arriving at a clinical diagnosis).
- Restoring the immune status was necessary to keep disseminated histoplasmosis in check (the patient has been free of illness for 8 years, even while on no antifungal treatment, because his CD4 count is adequate and his HIV viral load undetectable). This has been achieved by keeping him on HAART.
- Endogenous reactivation of a latent infection can occur with severe immunosuppression, as was shown in this patient.
- The microbiologic characteristics of this organism have been shown above, including the thermal dimorphic nature of this fungus (its behavior in tissue, and growth characteristics in various conditions and culture media).
- The purplish violaceous lesion in the hard palate resolved with antifungal treatment. It was not due to Kaposi's sarcoma, but represented a feature of disseminated histoplasmosis.

Reference

1 Baddley JW. Epidemiology and clinical manifestations of histoplasmosis in HIV-infected patients. Available at: www.uptodate.com/contents/epidemiology-and-clinical-manifestations-of-histoplasmosis-in-hiv-infected-patients?view=print, accessed March 2, 2016.

Further reading

1 Baddley JW. Diagnosis and treatment of histoplasmosis in HIV-infected patients. Available at: www.uptodate.com/online/content/topic.do?topicKey=pulm_inf/14229&selectedTitle=6%7E150&source=search_result, accessed March 2, 2016.

2 Casotti JA, Motta TQ, Ferreira Jr, CU, Cerruti Jr, C. Disseminated histoplasmosis in HIV positive patients in Espirito Santo State, Brazil: a clinical-laboratory study of 12 cases (1999–2001). Brazilian Journal of Infectious Diseases 2006; 10(5): 327–30.

3 Wheat J, Freifeld AG, Kleiman MB, et al. Clinical practice guidelines for the management of patients with histoplasmosis: 2007 update by the Infectious Diseases Society of America. Clinical Infectious Diseases 2007; 45(7):807–25.

4 Wheat J, Kauffman CA.Diagnosis and treatment of disseminated histoplasmosis in non-HIV-infected patients. Available at: www.uptodate.com/online/content/topic.do?topicKey=fung_inf/8533&view=print, accessed March 2, 2016.

Case 10.2   Recurrent multiorgan infection in an immunocompromised patient

A 40-year-old black male was brought to the ER in late May 2000 and admitted to the hospital because of lesions noted on brain scans done elsewhere.

Six days earlier, he had fallen at home following an observed seizure. He was taken to another hospital by the family, where the work-up included computed tomography (CT) scan and magnetic resonance imaging (MRI) of the brain. These studies suggested multiple cystic tumors. He was told that surgery was needed to make a definitive diagnosis, but he left out of panic, for fear of surgery, taking his x-rays with him. He said he wanted to be closer to his family.

He was subsequently brought to the ER by his family for further evaluation. There had been no previous seizure disorder. However, he complained of vague balance problems for an unspecified time, but was unable to give further details. He had no fever. He had lost 23 pounds in the past year, but related that to poor intake due to injury to his jaw.

His past medical history was significant for being physically assaulted 1 year earlier, leading to a broken jaw that required wiring by an oral surgeon. He claimed poor oral intake as a result of that injury. He had been unaware of any other serious medical illness until recently. He had no previous pneumonia, meningitis, diarrhea, or other systemic complaints. However, a few days before this admission, he had been told at the previous hospital that his HIV test was positive.

He was homosexual, with many previous sex partners over the years. He lived in a big cosmopolitan southern city in the 1980s, and only recently came home a few years ago. A former hotel help, he was currently unemployed. He was the middle child of a large family of seven children. He was not married, and had no children. He smoked half a pack of cigarettes a day for the past 20 years, and admitted to heavy alcohol intake (liquor and beer).

He was on no regular medications, and had no known drug allergies.

On 5/26/00, 1 day after admission, infectious diseases consult was sought to help address the cystic brain lesions in a patient with HIV infection.

Before my examination on 5/26/00, the patient had just had neurosurgery, a right posterior frontal burr hole with insertion of an external ventriculostomy catheter, designed to protect him against hydrocephalus, as well as to obtain cerebrospinal fluid (CSF) for studies.

On examination on 5/26/00, he was alert and oriented to place, but not to time and dates.

Vital signs showed the following: BP 158/98, RR 14, HR 74, temperature 97.7 °F (T_{max} 100 °F); height 5'9", and weight 146 pounds.

Head and neck exam showed a ventriculostomy drain in place, extensive oropharyngeal thrush, and generally poor dentition. The heart, lung, and abdominal exams were negative. The upper and lower extremities were unremarkable, and the skin showed no rash. External genital exam showed a normal uncircumcised male. Lymph nodes were palpable in the groin, shotty, multiple, and non-tender in the axilla, with smaller nodes noted in the right supraclavicular and epitrochlear areas. The neurologic exam found him to be awake and able to move all extremities; he had prompt deep tendon reflexes. Gait was not tested.

Chest x-ray showed patchy infiltrates in the right upper and left lower lobes. The CT scan and MRI of the brain done at the outside hospital on 5/20/00 showed multiple ring-enhancing lesions, one in the cerebellum, two or more in the right temporal lobe, and one in the left hemisphere. The largest lesion was in the cerebellum.

HIV serology was positive, but confirmatory Western blot test, CD4 count, and viral load were pending. CBC showed a WBC count of 7.4, H/H 11.1 and 32.0, respectively, MCV 101.6, and platelet count 278,000. Differential count showed 90% polymorphs, 1% band, 5% lymphs, and 4% monocytes. Sodium was 133, potassium 3.9, BUN and creatinine were 8 and 0.6 mg/dL, respectively. The liver tests showed the following: SGOT 82, SGPT 54 (normal), alkaline phosphatase 623, total bilirubin 1.23, albumin 3.1, and total protein 8.3. Cholesterol and triglyceride levels were within normal limits. Thyroid function tests (T3 uptake and T4) were normal. Electrocardiogram (EKG) was normal sinus rhythm, with sinus arrhythmia.

• *What is your presumptive diagnosis at this time?*

Hospital course

At surgery on 5/26/00, CSF fluid was obtained during placement of the ventriculostomy tube. It showed no WBCs, nine RBCs, normal glucose of 78 mg/dL, and protein of less than 10. Cryptococcal antigen test was negative. Additional studies included negative CSF VDRL test for syphilis. The CSF culture was negative.

One of the differential diagnoses considered to explain the multiple ring-enhancing brain lesions was toxoplasmosis. Serology for toxoplasma was, however, negative for IgG, IgM, and IgA antibodies. The patient was started on presumptive treatment for CNS toxoplasmosis, plus steroid therapy.

Over the next several days, he slowly developed increasing signs of obstructive hydrocephalus with lethargy, probably related to the cerebellar lesions. Serial CT scans and MRI suggested significant mass effect on the fourth ventricle, and reconfirmed multiple ring-enhancing lesions scattered in the cerebral hemispheres and cerebellum.

The previous CSF stains and cultures of 5/26/00 were still negative for bacteria, fungi, and acid-fast bacilli (AFB).

• *Have you changed or modified your differential diagnoses based on this new development?*

The patient was taken back to surgery on 6/7/00. A left posterior fossa craniectomy was done to reach the left cerebellar hemisphere. Four large abscesses with thick walls were encountered and evacuated. The pus was yellow brown, and did not have any detectable odor.

• *What do you expect the pus would show?*

The initial gram stain, KOH, and AFB stains of the brain abscess were negative. At 48 hours, however, an early growth was noted on the aerobic blood agar (BA) plate. The organism grew also on the TB culture medium (L J medium), but not on the fungal culture medium (SDA).

The BA plate with growth of the organism will be shown later.

The preliminary report at 2 days was 1+ filamentous gram-positive rods. Several days later, additional stains and tests could be done with more growth of the organism. The gram stain showed beaded, filamentous gram-positive rods while the modified acid-fast stain showed weakly, partial acid-fast filamentous rods. The organism was catalase positive and urea positive. Hydrolysis of casein and tyrosine was negative while xanthine hydrolysis was positive.

Additional tests performed by the laboratory later included confirmation of negative "equi factor" test (ruling out *Rhodococcus equi*) and antimicrobial sensitivity to vancomycin (34 mm disk), and resistance to erythromycin and gentamicin.

A non-standardized Kirby–Bauer (KB) antimicrobial susceptibility test for this organism was done, to help us tailor our treatment of the brain abscess. The organism showed sensitivity (by this KB test) to the following antimicrobial agents: trimethoprim/sulfamethoxazole, imipenem, ciprofloxacin, ceftriaxone, cefotaxime, amikacin, and vancomycin. Resistance was noted for the following agents: ampicillin, erythromycin, and gentamicin.

The patient was placed on high-dose intravenous trimethoprim/sulfamethoxazole (TMP-SMX), and eventually discharged on oral TMP-SMX on 6/20/00, to complete a planned prolonged therapy. Specific antiretroviral therapy was also to be started in the outpatient setting.

• *What is your new or revised diagnosis based on these laboratory findings, and the clinical decisions regarding the choice of therapy?*

Second admission (readmission) several months later

The patient was lost to follow-up after the last discharge from the hospital on 6/20/00.

He returned to the hospital on 2/10/01, 7.5 months after discharge. This time, he complained of a 3–4-week history of nausea, generalized malaise, cough with productive and purulent sputum, and sometimes vomiting, especially postprandial. He also had headache and left-sided chest pain. He had not been on antiretroviral medications, even though Medicaid had been approved for him a few months earlier. He had only learned about that fact recently. He had gained weight after being discharged in June, 2000, but had lost weight recently.

Two weeks earlier, he had been seen in the ER for "bronchitis" when he presented with cough and malaise. Fever and chills were prominent, with a temperature elevation up to 102 °F noted at home. Diarrhea had developed several days prior to admission (frequent, small, watery, dark stools). He complained of burning pain down his legs and feet on and off. In spite of a good appetite, he had lost 20 pounds in the last 4–5 weeks. He said he had quit smoking and drinking in the last weeks or months.

Following the first admission of 5/25–6/20/00, he had taken the prescribed TMP-SMX faithfully for 6 weeks. He then developed a severe generalized pruritic rash (exfoliative dermatitis), and so stopped the medication. He apparently was not on any alternative drug therapy for the previous brain abscess infection.

The patient haddeveloped a new drug allergy to TMP-SMX (as indicated), and to soma (unclear symptoms).

On 2/11/01, 1 day after readmission, infectious diseases consult was requested to address pneumonia and HIV infection in this patient.

On examination, he was alert and oriented, and in no acute distress. He had productive cough with yellowish brown sputum. His vital signs were as follows: BP 115/58, RR 22/min, HR 121, temperature 99°F; height 5'9", and weight 133.7 pounds. His head showed healed scalp scars, the mouth showed no thrush. Heart, lung, and abdomen were unremarkable on exam while the extremities were negative except for dry skin on the feet with peeling and athlete's foot. Lymph nodes were palpable but not enlarged in the neck, axilla, or groins. The neurologic exam showed him to be talkative and friendly; he had no focal findings.

The chest x-ray on admission showed bilateral pulmonary opacities in the right middle lobe, around the right hilum, as well as in the left upper mediastinum. There was no pleural effusion. This x-ray was clearly worse than the clear chest x-ray noted on 6/7/00, at the time of surgery for the brain abscess. CT scan of the chest done 2 days later on 2/12/01 confirmed dense consolidations, with cavitary disease noted in the left upper, right lower, and right middle lobes of the lung. Splenomegaly was also noted in the CT scan as an incidental finding. On the same day, MRI of the brain with contrast showed a focus of encephalomalacia within the left cerebellar hemisphere, likely representing an operative defect. No masses or other pathologic enhancements were noted. Generalized cortical atrophy was noted.

Blood cultures were negative. Chemistries showed a sodium value of 131, potassium 4.1, glucose 130 mg/dL, and BUN/creatinine 27/1.1, respectively. Liver panel showed SGOT 103, SGPT 45, alkaline phosphatase 128, albumin 3.1, total protein 8.1, and normal total bilirubin 1.09. The CBC showed a WBC count of 5.9, H/H of 5.7/16.5, respectively, MCV 93.2, and platelet count 353,000. The differential count showed 86% polymorphs, 8% lymphocytes, and 6% monocytes. Arterial blood gases (ABG) showed a pH of 7.49, pCO_2 31.3, pO_2 85, and O_2 saturation of 97% on room air.

Second hospital course

The working diagnosis was bilateral consolidating and cavitary pneumonia in an immune-compromised patient. The brain MRI suggested that the brain abscess was controlled, in spite of the patient having stopped treatment prematurely, on account of adverse drug reaction. Sputum stains and cultures were obtained.

While awaiting the sputum and blood culture results, empiric therapy was initiated to cover the likely organisms of pneumonia in this patient. Specific sputum culture was requested for *Nocardia*. *Pneumocystis jiroveci* stain was negative, as well as tests for *M. tuberculosis*. The patient had been started on a non-sulfa drug regimen, to provide coverage for PCP and bacterial pneumonia (clindamycin, levofloxacin, and primaquine). The clinical picture and chest x-ray findings were not typical for PCP. Adjustments in the antimicrobial regimen were made on 2/12/01, to provide coverage for the organism previously cultured from the brain abscess, as well as other bacterial causes of pneumonia. The patient was placed on ceftriaxone and imipenem.

The sputum gram stain suggested the presence of filamentous gram-positive rods, among other mouth flora. However, a few days later, the sputum culture was reported positive for dry, chalky, crumbly or sticky colonies on the BA plate. The sputum culture shown in Fig. 10.2a is similar to that from the brain abscess noted in June, 2000. The gram stain from this growth showed beaded filamentous gram-positive rods. This was similar in appearance to the brain abscess culture of 6/7/00.

A presumptive diagnosis of *Nocardia* pneumonia was made. Disseminated disease was suspected, although we did not document any site as positive other than the previous brain abscess. Intravenous ceftriaxone was continued, along with imipenem. The CD4 count was low at 52, and HIV-1 viral load was reported to be 13,266 copies /mL.

An in-house non-standardized KB susceptibility test for *Nocardia* this time showed a difference in the sensitivity pattern from June, 2000. This time, the organism was sensitive to tetracycline, imipenem, and ceftriaxone but resistant to TMP-SMX, ciprofloxacin, aztreonam, and ampicillin.

The patient improved in his general well-being while on treatment with intravenous ceftriaxone and imipenem, with reduction in cough symptoms. The fever, related to left forearm phlebitis, resolved. He was anxious to go home. At discharge home 16 days later on 2/26/01, he was placed on minocycline for a planned 6 months therapy, with outpatient follow-up. He was expected to begin antiretroviral therapy during the outpatient follow-up.

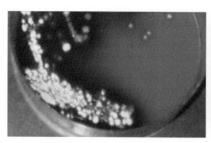

Figure 10.2a Aerobic BA plate: chalky white, crumbly or sticky colonies, more than 1 week old. (*See insert for color information*)

- *What is the unifying diagnosis in this patient?*
- *What is the likely prognosis with and without specific therapy?*
- *What was the organism cultured from the brain abscess and later from the sputum?*

Case discussion

The patient described here had acquired immune-deficiency syndrome (AIDS). By definition or classification, he falls under the old CDC surveillance classification of stage C-3 (C-III). This is because his CD4 count of 52 was below the cut-off point of 200 cells/mm^3. He also had the AIDS-defining illness of thrush, as well as disseminated opportunistic infection, apart from the systemic illnesses of anemia, severe weight loss or wasting, chronic diarrhea, etc. He was clearly very immune-compromised. The HIV infection was probably acquired many years before he presented with the seizure that ultimately led to the hospital admission.

The initial differential of his multiple cystic brain lesions included toxoplasmosis, especially because they were multiple, hypodense, and ring enhancing. Other differentials in this setting could also have included lymphoma, fungal or other abscesses, and metastatic disease.

Because of the danger of acute hydrocephalus, he had a ventriculostomy catheter placed on 5/26/00. He was also started on steroid therapy to address brain swelling (edema). However, his symptoms worsened while he still had the catheter in place, as well as on appropriate treatment for toxoplasmosis. He was at risk for obstructive hydrocephalus on account of the posterior fossa cerebellar mass lesions that threatened to compress the brainstem and fourth ventricle.

At surgery on 6/7/00, thick-walled abscesses were drained, providing needed relief for the patient. Once this was done, a correct diagnosis of *Nocardia* brain abscess was made, and he then received more specific and appropriate therapy with high-dose intravenous TMP-SMX. The initial treatment for presumed central nervous system (CNS) toxoplasmosis, sulfadiazine, also has activity against *Nocardia* that was subsequently confirmed.

He did well, and was later discharged on the oral form of TMP-SMX. He completed 6 weeks of this outpatient therapy before he developed a severe reaction to the TMP-SMX– exfoliative dermatitis with pruritus. He was lost to follow-up, and so received no alternative continuing therapy.

He returned 7.5 months later with consolidating and cavitating pneumonia. The brain CT scan and MRI confirmed clearance of the residual brain abscesses, but lung abscesses were now present. The in-house non-standardized sensitivity testing of the *Nocardia* suggested that the organism was sensitive to TMP-SMX. The fact that he felt well for several months even after stopping the drug, and the clearance of the residual brain abscesses attest to the correctness of this interpretation of the sensitivity report. Interestingly, the sensitivity of the pulmonary *Nocardia* in February of 2001, 8 months later, showed a change in the susceptibility of the same *Nocardia* species.

Our laboratory identified the *Nocardia* from both the brain abscess (June, 2000), and lung (February, 2001) as the same species: *Nocardia otitidiscaviarum* (formerly *N. caviae*). This organism was more resistant in February, 2001 than in June, 2000. In 2001, it was resistant to TMP-SMX, which we could not use, because of the new allergic reaction that developed before the 2001 admission. The patient responded well to the alternative drugs (ceftriaxone and imipenem) to which the organism showed in vitro sensitivity. In general, *N. otitidiscaviarum* shows more variable and unpredictable sensitivity than the more common *N. asteroides* complex or *N. asteroides* sensu strictu [1].

Nocardia belongs to a genus of aerobic actinomycetes responsible for localized or disseminated disease in animals and humans [1]. It is ubiquitous as an environmental saprophyte, present in soil, organic matter, and water. Human infections occur through direct inoculation of the skin or by inhalation [1]. It is not clear how this patient acquired his infection. It could have been through inhalation into the lungs or through the poor dentition. Either way, he developed disseminated disease involving the brain (abscesses) and lungs (consolidation and cavitary disease). It is also possible other organs were involved, although we cannot prove this with certainty. His liver tests were abnormal (acute hepatitis A, B, and C serologies were negative), and the spleen was enlarged. The blood cultures in this patient were negative, as has been typically reported for *Nocardia*, i.e. very rarely positive [1].

Although selective media can be used, *Nocardia* grows well in most non-selective media. It showed rapid growth in the aerobic blood agar, as well as in other media, including the mycobacteria media (L J medium). It did not grow in our fungal (SDA) media.

- *Final diagnosis: Disseminated Nocardia infection (brain abscess, pneumonia, and possibly other undetected systemic sites.*
- *N. otitidiscaviarum (N. caviae) was the likely isolate species incriminated.*

Postscript

Following the admission of 2/10–2/26/01, the patient was readmitted from 11/6 to 11/9/01 with fever, dry cough, and diarrhea. The chest x-ray was clear (no pneumonia); the diarrhea was thought to relate to one of the HIV meds (he had been on AZT/3TC and viracept, known to cause diarrhea).

By the next admission (2/12–3/1/02), he had been ill for several weeks with cough, fever, chills, weight loss, nausea, and vomiting. He had not taken his HIV meds for months. His chest x-ray showed pneumonia (diffuse interstitial infiltrates). New HIV meds were restarted. He was watched for 8 days to assure his tolerance of the medications before discharge. His CD4 on 2/13/02 was the lowest it had ever been, 4, while the viral load was 219,239 copies /mL.

During the first office visit, on 3/11/02, anorexia, elevated amylase and lipase levels suggested pancreatitis. When seen again on 3/20/02, clinical symptoms and laboratory parameters were improved after adjustment in HIV meds (fever and diarrhea resolved; liver and pancreas enzyme levels improved towards normal).

However, peripheral neuropathy symptoms (difficulty walking, burning of feet) persisted in April, 2002. The patient was admitted to the hospital in mid-April with cachexia, mucositis, and bilateral pneumonia. *Pseudomonas aeruginosa* was cultured from the sputum. Tests for PCP, AFB, and *Nocardia* were negative. Cytomegalovirus (CMV) serology was positive, but the interpretation was unclear. Blood cultures were negative. He had stopped all HIV meds since March, 2002 because he could not tolerate them.

In May, 2002, he was placed under hospice for comfort care.

He was readmitted to the hospital twice in July and once for the last time in August, 2002. His pneumonia progressed to respiratory failure. He remained cachectic, with unresolved pneumonia. He died in early August, 2002, probably from end-stage AIDS and respiratory failure.

Final comments

This patient had many prognostic factors against him. He presented late at a time when he was already severely immunosuppressed with advanced HIV infection (AIDS). Once a diagnosis was made, he was provided with appropriate medication (TMP-SMX) to which he responded. However, that medication was stopped because of a serious adverse drug reaction. He went for several months without taking needed medication for the *Nocardia* or appropriate alternative drug treatment. He failed to keep scheduled outpatient follow-up appointments. An outside physician placed him on antiretroviral therapy that he was unable to tolerate, so he stopped. When these were restarted or replaced, he did well for a while, but then developed pancreatitis to the HIV drug(s). HIV drug therapy was also started late because HIV infection was diagnosed late into his illness, after which there was delay in starting HIV-specific therapy because it took several months to obtain Medicaid insurance . In short, he did not have HIV-specific medications initially, and when he did, he either could not tolerate or did not take them. He developed multiple complications of his HIV disease: complications of treatment of the HIV, as well as complications of treatment of the opportunistic infection(s).

Lessons learned from this case

- Successful treatment of an opportunistic infection requires that therapy be started early and maintained consistently. Neither was the case with the patient in this report.
- Early diagnosis is a prerequisite for starting treatment early, and a chance for control of the disease. In this patient, both the HIV infection and the opportunistic infection were diagnosed at late stages.

- Social issues like access and affordability to care did have an impact on the poor outcome of the illness in this patient. He was unable to obtain health insurance early in his illness.
- Without control of the severe immune deficiency, even appropriate therapy can be ineffective. This patient was unable to tolerate his HIV-specific drugs, and so he remained severely immune-compromised all through his illness.
- Disseminated nocardiosis remains a difficult disease to treat even in the best of circumstances.

Reference

1 Sorrell TC, Mitchell DH, Iredell JR. *Nocardia* species. In: Mandell GL, Bennett JE, Dolin R (eds) *Principles and Practice of Infectious Diseases*, 6th edn. Philadelphia: Churchill Livingstone, 2005, pp. 2916–24.

Further reading

1 Case records of the Massachusetts General Hospital. Weekly clinicopathological exercises. Case 29-2000. A 69-year old renal transplant recipient with low grade fever and multiple pulmonary nodules. New England Journal of Medicine 2000; 343(12):870–7.

2 Gallant JE, Ko AH. Cavitary pulmonary lesions in patients infected with human immunodeficiency virus. Clinical Infectious Diseases 1996; 22(4): 671–82.

3 Mathisen GE, Johnson JP. Brain abscess. Clinical Infectious Diseases 1997; 25(4): 763–79.

4 Menéndez R, Cordero PJ, Santos M, Gobernado M, Marco V. Pulmonary infection with *Nocardia* species: a report of 10 cases and review. European Respiratory Journal 1997; 10(7): 1542–6.

5 Naguib MT, Fine DP. Brain abscess due to *Nocardia brasiliensis* hematogenously spread from a pulmonary infection. Clinical Infectious Diseases 1995; 21(2): 450–60.

6 Roberts SA, Franklin JC, Mijch A, Spelman D. *Nocardia* infection in heart-lung transplant recipients at Alfred Hospital, Melbourne, Australia, 1989–1998. Clinical Infectious Diseases 2000; 31(4): 968–72.

7 Uttamchandani RB, Daikos GL, Reyes RR, et al. Nocardiosis in 30 patients with advanced human immunodeficiency virus infection: clinical features and outcome. Clinical Infectious Diseases 1994; 18(3): 348–53.

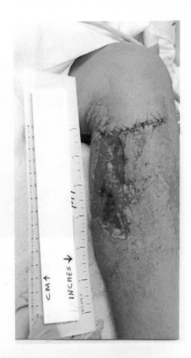

Figure 1.1a Left calf wound on day 4 post surgery (reproduced with permission).

Figure 1.1b Luxuriant growth of gram-negative rod on blood agar medium in 24 hours, and comparative growth of the same organism on BA, chocolate BA, and MacConkey agar.

Cases in Clinical Infectious Disease Practice: Obtaining a Good History from the Patient Remains the Cornerstone of an Accurate Clinical Diagnosis: Lessons Learned in Many Years of Clinical Practice,
First Edition. Okechukwu Ekenna.
© 2016 John Wiley & Sons, Inc. Published 2016 by John Wiley & Sons, Inc.

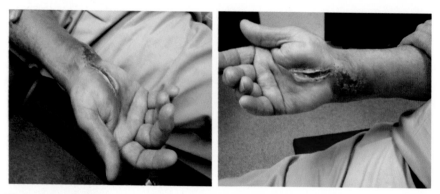

Figure 1.2a A large bulge of purplish bruised area is noted in the right wrist, along with the surgical incision in the hand. Photo was taken on 3/4/09 (reproduced with permission).

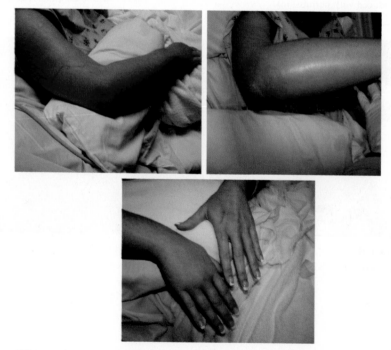

Figure 1.3a Swelling of the right arm, elbow, and hand, reflecting cellulitis. Compare the left hand without swelling. Photos taken on 1/6/10 (reproduced with permission).

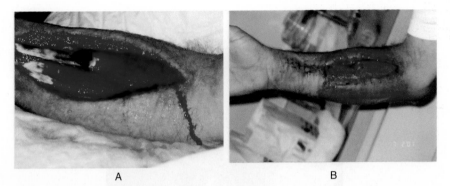

Figure 1.6a Right forearm: A. Day 15 post admission, after several debridements (note necrotic tendon). B. Day 21 of hospitalization: initial plastic surgery. Photos were taken on 6/26 (A) and 7/2/01 (B) (reproduced with permission).

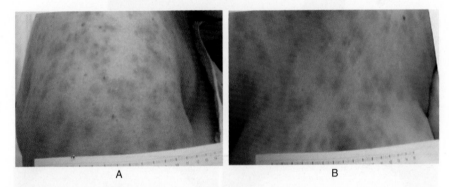

Figure 3.3a A. Right upper arm and shoulder: maculonodular lesions, some scaling noted. B. Anterior chest lesions: maculonodular, only partially blanching. Photos taken on 1/13/11 (reproduced with permission).

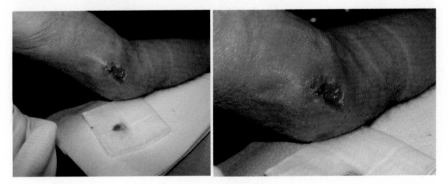

Figure 3.4a Left arm and elbow edema with inflammatory abscess: note the bluish area of the wound with abscess at the edge. Office photo taken on 3/18/09 (reproduced with permission).

Figure 4.1a Proglottid of tapeworm obtained from the 13-month-old Caucasian male (March, 2006).

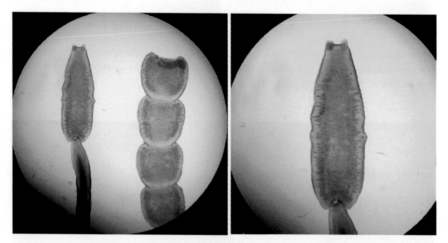

Figure 4.1b These digital images depict tapeworm proglottids from our patient: the mature proglottid on the right shows lateral genital pores and "rice-grain" shape, while the immature proglottids are broader in appearance. Courtesy of Henry S. Bishop, CDC, Atlanta.

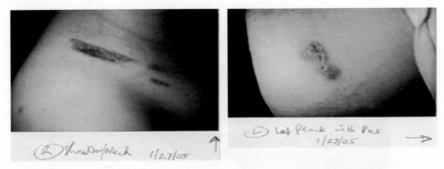

Figure 4.2a Right shoulder/neck and left lateral flank areas with tiny tunneling abscesses (reproduced with permission).

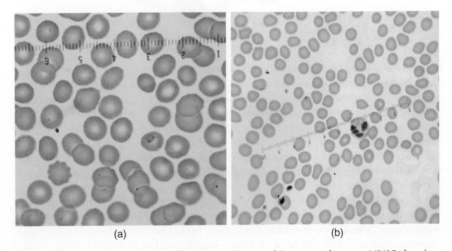

(a)   (b)

Figure 5.3a A. Peripheral blood malaria stain (Giemsa, thin smear) done on 6/9/15 showing intracellular ring forms in normal-sized red cells. B. Segmented PMN noted for size comparison.

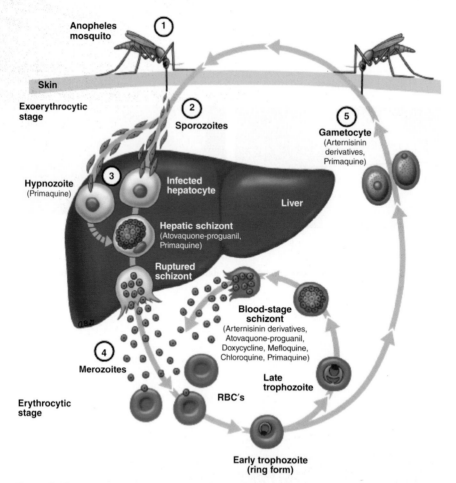

Figure 5.4b Life cycle of *Plasmodium*. (1) *Plasmodium*-infected *Anopheles* mosquito bites a human and transmits sporozoites into the bloodstream. (2) Sporozoites migrate through the blood to the liver where they invade hepatocytes and divide to form multinucleated schizonts (pre-erythrocytic stage). Atovaquone-proguanil and primaquine have activity against hepatic-stage schizonts. (3) Hypnozoites are a quiescent stage in the liver that exist only in the setting of *P. vivax* and *P. ovale* infection. This liver stage does not cause clinical symptoms, but with reactivation and release into the circulation, late-onset or relapsed disease can occur up to many months after initial infection. Primaquine is active against the quiescent hypnozoites of *P. vivax* and *P. ovale*. (4) The schizonts rupture and release merozoites into the circulation where they invade red blood cells. Within red cells, merozoites mature from ring forms to trophozoites to multinucleated schizonts (erythrocytic stage). Blood-stage schizonticides such as artemisinins, atovaquone-proguanil, doxycycline, mefloquine, and chloroquine interrupt schizogony within red cells. (5) Some merozoites differentiate into male or female gametocytes. These cells are ingested by the *Anopheles* mosquito and mature in the midgut, where sporozoites develop and migrate to the salivary glands of the mosquito. The mosquito completes the cycle of transmission by biting another host.

*There is strong evidence that drugs listed in parentheses are active against designated stage of parasitic life cycle. Primaquine is a blood-stage schizonticide with activity against schizonts of *P. vivax* but not those of *P. falciparum*.

•Reproduced with permission of UpToDate, Inc. from Hopkins H. Diagnosis of malaria. Available at: http://www.uptodate.com/contents/diagnosis-of-malaria?source=search_result&=Malaria&selectedTitle=4%7E150, accessed February 26, 2016.

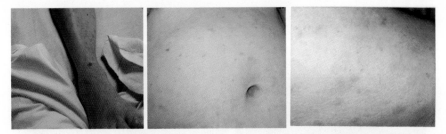

Figure 6.1a Tick bite eschar in anterior right ankle, and macular papular rash on abdominal torso and upper back (reproduced with permission of the Mississippi State Medical Association).

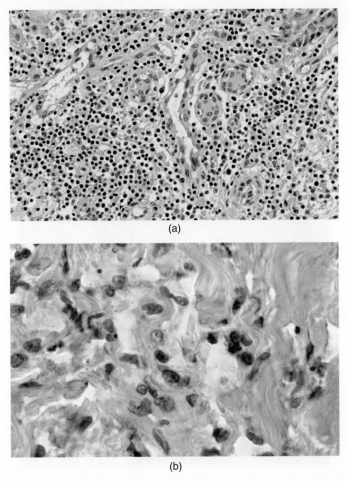

(a)

(b)

Figure 6.2b Histopathology and immunohistochemical staining for spotted fever group rickettsiae (SFGR) of right groin eschar of case #5. A. Hematoxylin and eosin stain showing diffuse lymphocytic perivascular inflammation: original magnification ×100. B. Immunohistochemical stain for SFGR (red). Original magnification ×258 (reproduced with permission of the Mississippi State Medical Association).

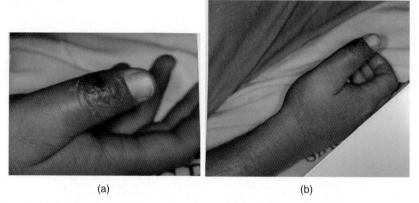

(a)   (b)

Figure 7.1a A. Right thumb: scabbed pustular area with skin peeling. B. Inflammation involving both the thumb and radial dorsal ball of the hand.

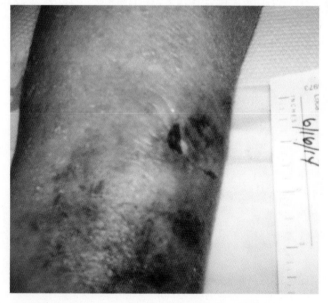

Figure 7.2a Photo taken on the day of admission, 6/16/14, 1 day post laceration of right medial mid-distal shin. Size of blister bruise was noted the next day to be 12.5 × 15 × 0.2 cm (reproduced with permission).

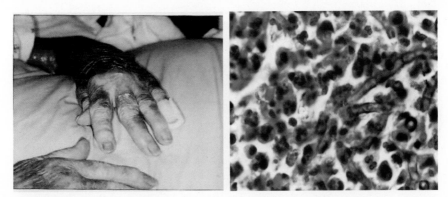

Figure 8.1b New keratotic lesion on third left finger (photo taken 2/28/05), and histopathology of the biopsy (H&E stain) done on 3/3/05. Note the giant cells and the large brown, thick-walled, septate hyphae (reproduced with permission).

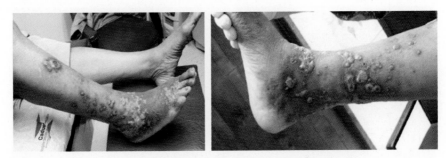

Figure 8.2c Progressive chromomycosis of the right leg, 16 years after onset. Photo taken on 5/29/09 (reproduced with permission).

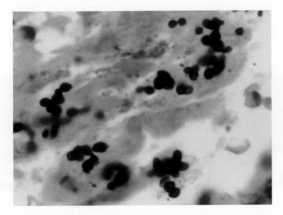

Figure 10.1b GMS stain of RLL transbronchial biopsy done on 1/11/07. Magnification ×40: yeast forms, average size 2.5 microns.

Figure 10.1f 21-day-old mold culture on brain heart infusion (BHI) and Sabouraud dextrose agar (SDA).

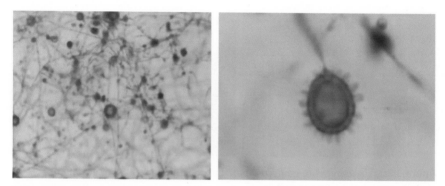

Figure 10.1g Slide/tease mount of 21-day-old culture with the larger macroconidia and the smaller microconidia evident in this mycelial phase *Histoplasma*; the macroconidium or tuberculate form with protrusions is also shown in the next picture. The macroconidia are usually 8–15 microns in size, while the microconidia are about 2–5 microns.

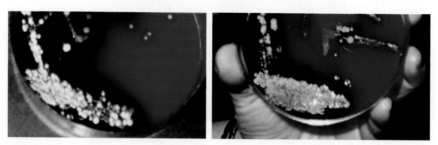

Figure 10.2a Aerobic BA plate: chalky white, crumbly or sticky colonies, more than 1 week old.

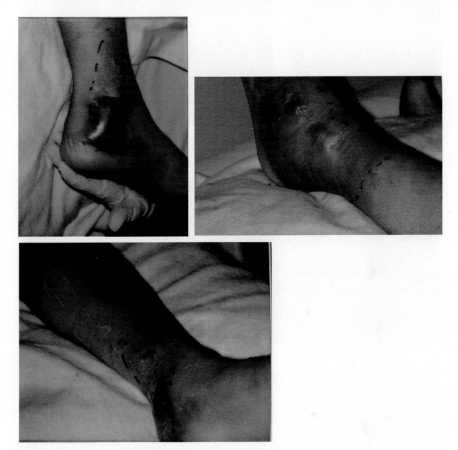

Figure 12.1a Hemorrhagic blisters noted at the medial and lateral left foot, ankle and heel. Photos taken on 9/2/05 (reproduced with permission).

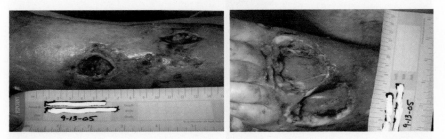

Figure 12.2b Necrotizing fasciitis left shin and foot, 6 days after initial debridement. Photo taken 9/13/05 (reproduced with permission).

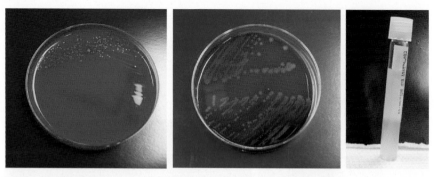

Figure 13.2a Blood agar plate cultures: A. Day 1. B. Day 9. C. Broth culture of CSF on day 9.

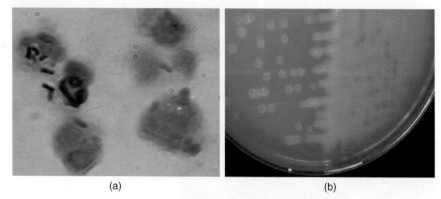

(a)　　　　　　　　　　　　　　(b)

Figure 13.2b A. Gram stain of cerebrospinal fluid showing inflammatory cells and small, gram-positive rods and coccobacilli. B. Clear areas around each colony of *L. monocytogenes* are characteristic small zones of beta-hemolysis on blood agar [1]. Reproduced with permission from: Waltzman M. Initial evaluation of shock in children. Available at: http://www.uptodate.com/contents/initial-evaluation-of-shock-in-children, accessed on March 4, 2016. Copyright ©2015 UpToDate, Inc. For more information visit http://www.uptodate.com.

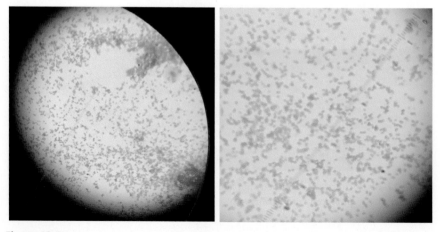

Figure 13.4c Gram stain of 2-day-old growth on CBA on 9/9/15: gram-negative diplococci and coccobacillary forms.

CHAPTER 11

Toxic Manifestations of Infectious and Non-infectious Diseases

Case 11.1   An odd presentation for toxic shock syndrome

A 46-year-old Caucasian female was admitted in July 2007 with nausea, vomiting, diarrhea (up to 20 bowel movements in 9 hours), and rash. The illness was acute in onset, with presentation to the hospital emergency room (ER) within 7 hours of symptom onset. In the ER, she was found to have a blood pressure (BP) of 74/44, pulse of 112, and a temperature of 99.8 °F. She received large volumes of intravenous fluid, and was admitted to the floor.

While on the floor, her BP continued to stay low, with systolic hypotension in the 70s and 60s, and a heart rate between 120 and 150. She was thus transferred to the medical intensive care unit (MICU) the next day where she was placed on pressor support therapy, in addition to intravenous fluids.

Two days after admission, infectious diseases consult was sought for septic shock, diarrhea, fever, and rash. She was seen in the MICU.

The review of systems was significant for generalized body aches and muscle tenderness, but no headaches or meningismus. She had watery diarrhea, with up to 20 bowel movements in 8–9 hours. She was orthostatic in the ER, with BP of 103/54 and pulse of 100, dropping to 70/51, with a pulse of 112 from lying to standing position. The throat was dry, but not sore. Even 2 days after admission, her stool was still watery/loose and greenish in color.

Family and social history: she was married with two young children, and worked as a nurse caring for disabled adults. She did not smoke or drink alcohol. She had allergies to codeine (itch) and non-steroidal anti-inflammatory drugs (shortness of breath or exacerbation of asthma symptoms).

Epidemiologic history: there was no outbreak of diarrhea or acute illness at home or at work. One of the children had some minor sore throat complaints, but no active illness. Their dog and cat were healthy. There had not been any recent travels. She denied active menses at the time of evaluation.

The past medical history was significant for tubal ligation 5 years earlier, sinus surgery 7–8 years ago, a history of depression, ovarian cysts, and stable asthma for 20 years.

Cases in Clinical Infectious Disease Practice: Obtaining a Good History from the Patient Remains the Cornerstone of an Accurate Clinical Diagnosis: Lessons Learned in Many Years of Clinical Practice,
First Edition. Okechukwu Ekenna.
© 2016 John Wiley & Sons, Inc. Published 2016 by John Wiley & Sons, Inc.

Her medications prior to admission included venlafaxine (for depression) 75 mg/day, and fluticasone/salmeterol (for asthma) 250/50 2×/day.

On examination in the MICU, she looked moderately ill. Her vital signs were as follows: BP 98/62, respiratory rate (RR) 38, heart rate (HR) 118, temperature 100.1 °F (maximum temperature 103.1 °F the day before); height 5'4", weight 128 pounds.

Head and neck exam showed extremely dry mucosa and tongue; heart and lung exam was unremarkable. Abdomen was soft with slight generalized soreness, but no organomegaly; bowel sounds were diminished; she still had greenish mucoid stools. Lymph nodes were not enlarged; upper and lower extremities showed non-pitting edema. The skin showed a generalized erythematous flush of the face and back, with blanching on pressure or palpation. Several blisters were noted on the inner thighs and upper left flank. There were no lesions on the palms or soles. The neurologic exam showed her to be somber and depressed, but appropriate. She had no signs of encephalopathy or meningeal signs, but was aching in all body muscle groups and areas.

Laboratory and x-ray findings included the following: chest x-ray was clear, while the computed tomography (CT) scan of the abdomen and pelvis showed marked thickening of the walls of the stomach, and mildly so in the duodenum. The electrocardiogram (EKG) showed a sinus tachycardia of 149 beats per minute. The complete blood count (CBC) on admission showed a white blood cell (WBC) count of 10.1, hemoglobin and hematocrit (H/H) 14.3/41.4, respectively, mean corpuscular volume (MCV) 85.6, and platelets 198,000. Sodium was 134 mEq/L, potassium low at 2.5, glucose 145 mg/dL, and blood urea nitrogen (BUN)/creatinine 18/1.0, respectively. The next day sodium was down to 125 and BUN/creatinine 34/2.5, an estimated glomerular filtration rate of 27 mL/min. Stool culture was positive for methicillin-sensitive *Staph. aureus* (reported later).

Hospital course

On the day of the ID consult, the patient was on hydrocortisone sodium intravenously. WBC count was 17.6, with a differential count that showed 34% polymorphs, 35% bands, 3% lymphocytes, 20% metamyelocytes, and 6% myelocytes. The liver function tests showed mild transaminase elevation (serum glutamic oxaloacetic transaminase/serum glutamic pyruvic transaminase [SGOT/SGPT] were 168/58, respectively), but with normal bilirubin and low albumin of 2.5 mg/dL. Creatine phosphokinase (CPK) was elevated (5090) 2 days after admission, and by the next day was down to 2358.

A suboptimal vaginal exam was done by the primary care physician showing nothing significant. The vaginal cultures were later reported positive for *E. coli* and group B *Streptococcus*. One of two blood cultures drawn 1 day after admission was positive for *Enterobacter agglomerans* (clinically this was not considered

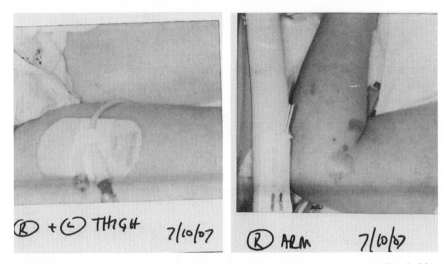

Figure 11.1a Erythema and tiny blisters in between the thighs, and erythema with Nikolski sign in the right arm (photos taken 2 days after ID consult, 4 days after admission) (reproduced with permission).

significant, but likely from an enteric source – stool contamination). Cultures of the inner thigh blisters were negative for pathogens.

On 7/10/10 (4 days after admission, 2 days after ID consult), examination of the skin showed some spreading blistering consistent with Nikolski sign and suggestive of toxic epidermal necrolysis (TEN) or toxic shock syndrome (TSS). On that day, the platelet count dropped to 36,000. Because of concern for TSS even after 4 days of antibiotic therapy and hemodynamic support, the ID consultant suggested a change in therapy as follows: intravenous immunoglobulin (IgG) was started, along with ceftriaxone, while clindamycin started 2 days earlier was continued, along with levofloxacin. Metronidazole, vancomycin, and pipracillin/tazobactam were discontinued.

Photographs of the skin were taken (Fig. 11.1a).

Further clinical and laboratory developments

Ten days after admission, the patient had no new skin blisters; the old ones had dried up and open areas were now scabbing over. Edema was generally down and erythematous rash was resolved. Thrombocytopenia was resolved (platelets 279,000), WBC count was normal (8.7), and BUN/creatinine were normal (9/0.6, respectively). Peripheral intravenous access was established and central line discontinued. Antimicrobials were changed to oral: clindamycin and levofloxacin.

The patient had been moved to the floor, and was now allowed to shower for the first time since admission (she had been in the MICU and too ill previously).
- *What is your diagnosis?*

TAMPON   7/16/07

Figure 11.1b Intravaginal tampon that fell out during showering 10 days after admission (reproduced with permission).

Our diagnosis was toxic shock-like syndrome, although we had no definitive proof.

Surprise supporting evidence for diagnosis

While taking a shower on hospital day 10, the patient found a tampon in her vagina (the tampon fell out during the shower!). She had been too ill to recall that she had a tampon in place before she became ill. Previous suboptimal vaginal exam on 7/8/07, 8 days earlier, had failed to identify the tampon.

A photograph of the intravaginal tampon that fell out of the patient was taken (Fig. 11.1b).

- *Final diagnosis: Toxic shock syndrome, most likely related to the intravaginal tampon.*

Case discussion

This patient was very ill at presentation, and had all the hallmarks of toxicity. She had no memory that she had placed a tampon in her vagina prior to admission, or exactly when she did that. She was not menstruating at the time of evaluation in the hospital. The vaginal exam performed in the MICU was suboptimal, and was not a full speculum exam. The tampon was therefore missed. This could have had more serious consequences (not removing the tampon immediately). Clinically, her symptoms were consistent with toxic shock syndrome and she was treated as such, except that the tampon was not removed in a timely fashion.

I suspect that the *Enterobacter agglomerans* cultured in one of the two blood cultures was likely unrelated to her critical illness, but more likely a contaminant from the gastrointestinal (GI) tract. She had profuse diarrhea reminiscent of cholera, but *Vibrio cholerae* was not cultured from the stool. *Staph. aureus* (methicillin sensitive) was cultured from the stool and not from the blood. This is not a requirement for the diagnosis of staphylococcal TSS. The severe diarrhea is part of the syndrome of toxic shock.

Toxic shock syndrome was most likely related to (or caused by) staphylococcal TSS toxin-1 (TSST-1) or other exotoxins. This is a speculation on our part since these toxins were not assayed. The deep intravaginal tampon would be compatible with the menstrual form of TSS. As mentioned, *Staphylococus aureus* was cultured from the stool 1 day after admission, but not from the blood.

Clinical support for the diagnosis of toxic shock syndrome includes the following (Box 11.1a).

Box 11.1a Case definition for toxic shock syndrome (TSS) from the Centers for Disease Control [1,2]

Fever

- T >38.9°C (102.0 °F)

Hypotension

- Systolic blood pressure ≤90 mmHg for adults or <5th percentile by age for children <16 years of age; orthostatic drop in diastolic blood pressure ≥15 mmHg
- Orthostatic syncope or dizziness

Rash

- Diffuse macular erythroderma
- Desquamation: 1–2 weeks after onset of illness, particularly involving palms and soles

Multisystem involvement (three or more of the following organ systems)

- GI: Vomiting or diarrhea at onset of illness
- Muscular: Severe myalgia or CPK elevation >2 times the normal upper limit
- Mucous membranes: Vaginal, oropharyngeal, or conjunctival hyperemia
- Renal: BUN or serum creatinine >2 times the normal upper limit, or pyuria (>5 WBC/hpf)
- Hepatic: Bilirubin or transaminases >2 times the normal upper limit
- Hematologic: Platelets <100,000/µL
- Central nervous system: Disorientation or alterations in consciousness without focal neurologic signs in the absence of fever and hypotension

---

- Profuse diarrhea (20 bowel movements in 8–9 hours)
- Stool character supports toxin-induced diarrhea (no WBCs, negative *C. difficile* toxin, cholera-like)
- Septic shock presentation (hypotension, dehydration, renal insufficiency, liver function abnormality)
- Severe edema of skin and gastrointestinal wall
- Rash, blisters, Nikolski sign
- Multiorgan system involvement (GI tract, renal, liver, skin)

Types or forms of TSS

There are menstrual and non-menstrual types of TSS [2].

The proportion of TSS syndromes related to menstruation has decreased over the past 20–30 years. Some of this decrease is likely related to withdrawal of highly absorbent tampons and polyacrylate rayon-containing products from the market. However, tampon use remains a risk factor for TSS. Women who develop TSS are more likely to have used tampons with higher absorbencies, used tampons continuously for more days of their cycle, and kept a single tampon in place for a longer period of time. This was the case in our patient reported here.

Approximately one-half of reported TSS cases are non-menstrual. Non-menstrual TSS has been seen in a variety of clinical situations, including surgical and postpartum wound infections, mastitis, septorhinoplasty, sinusitis, osteomyelitis, arthritis, burns, cutaneous and subcutaneous lesions [2].

Lessons learned from this case

- This infection had all the hallmarks of a toxin-induced illness.
- The clinical entity of TSS syndrome was recognized but not pursued to the logical conclusion (not finding the tampon early in the illness).
- The patient's memory could not be relied upon regarding her menstruation history.
- The pelvic examination should have been pursued with a full speculum exam (suboptimal exam in the MICU in this very ill patient may sometimes require consultation with the gynecologist).
- Highly absorbent tampons (in place for several days) present the greatest danger for menstrual cases of TSS.
- Keep in mind that half the cases of TSS are non-menstrual.
- Aggressive therapy for TSS probably led to a good outcome in this patient, in spite of the delay in removing the offending tampon.

References

1 CDC. Case definitions for infectious conditions under public health surveillance. Morbidity and Mortality Weekly Report 1997; 46(RR-10): 38–40.
2 Chu VH. Staphylococcal toxic shock syndrome. Available at: www.uptodate.com/online/content/topic.do?topicKey=gram_pos/4986&view=print, accessed March 3, 2016.

Case 11.2   A peculiar and dramatic presentation of septic shock

In early June, 2010, an 18-year-old Caucasian male was admitted with the chief complaint of "feel like I am going to die."

Seven days earlier, he had developed sore throat and fever (temperature 101 °F), and had received treatment at an urgent care center with azithromycin. Two days after the 5-day antibiotic therapy finished, he developed new and more severe symptoms that included nausea, vomiting, diarrhea, and abdominal pain. Thirty six hours after the onset of these new symptoms, the ambulance was called and he was brought to the ER in shock. He had no fever at the time.

The review of systems was significant for the following.

- One week prior to admission (PTA) he had fever and sore throat.
- Thirty-six hours PTA: nausea, vomiting, diarrhea, abdominal pain, and posterior headache.
- Abdominal pain was very severe (10/10 in pain intensity).
- Nausea and vomiting: up to 15× in 24 hours.
- Diarrhea: up to 20 loose stools in 24 hours.
- On the day of admission: severe weakness and "afraid I was not going to make it."

On presentation to the ER, he appeared very ill and toxic. His initial vital signs showed a BP of 82/37, pulse 120, RR 22, temperature 97.4 °F. Even after 3 liters of normal saline, the BP had dropped further to 63/28. He had copious brownish liquid stool. A diagnosis of acute sepsis and septic shock was made, and he was admitted to the ICU.

Epidemiologic history: a recent high school graduate living out of state, he had come home 2 weeks earlier (1 week before onset of sore throat). A cousin with whom he shared space and time together had "strep throat" a few days before he became ill, and was given a penicillin-like drug. The cousin recovered. His parents were not ill with any respiratory illness. In the 2 weeks PTA he had spent a lot of time outdoors in the heat of the sun doing manual labor.

Family and social history: he was the younger of two brothers, had just graduated from high school and was looking forward to going to college. There was no history of tobacco or alcohol abuse. He had lived until recently on the West Coast of the USA. There was no known drug allergy.

The past medical history was significant for a history of depression. He had been on some antidepressant for years and was being weaned off sertraline (over weeks) at the time of admission. He also presumably had a previous diagnosis of attention deficit hyperactivity disorder (ADHD), and was on methylphenidate. There were no other serious medical conditions known.

Hospital course

In the ER, the patient complained of severe, sharp, constant, and pressure-type abdominal pain described as being 10/10 in intensity. Following x-ray studies, including CT scan of the abdomen, the surgical consultant took the patient to the operating room urgently for exploratory laparotomy. The preoperative diagnosis was diffuse peritonitis.

At surgery, about 2 liters of non-feculent, thick yellowish brown pus was found and cleaned out of the abdomen. The gallbladder (not diseased) was removed, to better visualize the duodenum (Kocher maneuver). No enterotomies or perforation were noted at surgery.

The patient stayed in the ICU for over 1 week. He was hospitalized for a total of 16 days. While hospitalized, he was seen by multiple consultants (in addition to the attending physician) to address the myriad organ dysfunctions and problems that developed.

- General surgery (6/9/10): for exploratory laparotomy for peritonitis.
- Urology (6/9/10): to watch and follow because of abnormal urinalysis and reduced urine output initially. No procedures were performed.
- Pulmonary/critical care: for ICU care, ventilator and general management.
- Cardiology (6/11/10): for elevated cardiac enzymes, premature ventricular contractions, myocarditis, and low cardiac ejection fraction.
- Gastroenterology (6/10/10): because of initial nausea, vomiting, diarrhea, and abdominal pain.
- Nephrology (6/11/10): for acute renal failure.

Thirteen days after admission, infectious diseases consultation was requested to address issues of multiorgan system dysfunction or failure, attempt a unifying diagnosis, and address the duration of antimicrobial therapy going forward.

On examination on the day of admission, the patient had looked toxic and was hypotensive even after 3 liters of normal saline in the ER: BP 63/28, RR 22, HR 120, temperature 97.4 °F; height 6'2", weight 215.8 pounds.

On the day of the ID consult (13 days after admission), he was in no acute distress. He was now on the general medical floor and out of the ICU. The BP was 128/71, RR 18, HR 84, temperature 96.6 °F; weight 201.7 pounds (down by 14 pounds). The head and neck exam showed right subconjunctival hemorrhage, no adenopathy, and clear throat. The heart and lung exam was unremarkable. The abdomen showed surgical sites in the mid and right side; the drains had been pulled, and there was no purulence noted. The external genitals showed normal circumcised male phallus; there was tinea cruris of the groin, and no adenopathy. The skin showed a good suntan and freckles, no rash, but some exfoliation, including of the axilla. Neurologic exam was normal. The extremities showed extensive exfoliation of the hands and feet (Fig. 11.2a).

- *What is the likely diagnosis?*

The radiologic findings showed the following: the chest x-ray on admission showed no definite pneumonia; the CT scan of the abdomen showed some stranding within the retroperitoneal area consistent with mild inflammation. Repeat chest x-ray 10 days later showed the heart to be mildly enlarged, but the lungs remained clear. CT scan of the abdomen and pelvis 9 days after admission was described as showing edema of the abdominal wall, and likely retroperitoneal adenopathy, and possibly small splenic infarct. Some extensive

Figure 11.2a Sketch of the feet on day 13 of admission (the day of ID consultation). Exfoliation of feet. The shaded areas reflect a thick, peeling, horny layer while the unshaded areas reflect exfoliated areas with exposed erythematous skin (sketch adapted from a photograph taken on 6/22/10).

myocardial calcification within the heart ventricles was also reported, along with moderate bilateral pleural effusions.

Laboratory findings included the following: on admission, sodium level was 134 mEq/L, potassium 3.1, glucose 102 mg/dl, BUN/creatinine 20/3.7, respectively; albumin was low at 2.6 mg/dl. Lumbar puncture done to rule out meningitis was negative for cells and other parameters, including a negative culture. CBC on admission showed a WBC count of 36,700/μL, H/H 12/34, MCV 83.3, platelet count 219,000. The differential count showed 53% polymorphs, 45% bands, 1% eosinophils, and 2% metamyelocytes. Stool studies for *C. difficile* toxin were negative on admission and 6 days later, although the stool WBC was found to be elevated on the day of admission (>30/hpf). The urinalysis showed >30 red blood cells and WBCs, but negative cultures on two occasions. HIV serology was negative. The disseminated intravascular coagulation (DIC) panel 1 day after admission was abnormal with prothrombin time of 23.8 seconds, partial thromboplastin time 45 seconds, with platelets 103,000; D-dimer was elevated at 9.96 (normal <0.48 μg/mL). On day 2 of admission, the CPK was high at 2812 ng/mL, troponin I was 30.45 (normal <0.05), creatine kinase myocardial band (CK-MB) 293.5 ng/mL (N<5). Renal function worsened, with

BUN/creatinine 44/4.9, respectively; lactate dehydrogenase (LDH) was 545, SGOT 296, SGPT 89, alkaline phosphatase 110 (normal), total bilirubin 4.9, total protein 4.9 mg/dL, and albumin 2.1 mg/dL. WBC count remained high at 33,200/μL, and platelet count dropped to 81,000/μL.

Echocardiogram on day 2 of admission showed an ejection fraction (EF) of 20%, with global hypokinesis, consistent with myocarditis; 10 days later EF had improved to 35%. Transesophageal echocardiogram was negative for vegetations.

Peritoneal and blood cultures on admission were positive with the same organism.

- *Have you formulated a new or different diagnosis?*
- *What organism do you think was found in the blood and peritoneum?*

Case discussion

Group A *Streptococcus* (GAS), also called *Streptococcus pyogenes*, was cultured from both blood and peritoneum in this patient.

To begin the discussion, let us summarize the major findings and complications experienced by this patient.

- Initial sore throat and fever
- Severe diarrhea, nausea/vomiting, and abdominal pain
- Severe hypotension (septic shock)
- Peritonitis
- Leukocytosis (WBC up to 36,700)
- Acute renal failure
- Myocarditis (elevation of cardiac enzymes, EKG changes, and low EF)
- Liver test abnormality
- Elements of DIC
- Exfoliation of skin of hands, feet, and axilla
- Positive blood and peritoneal cultures for GAS

This very ill young man was discharged home to outpatient follow-up 16 days after admission, and 4 days after the ID consult. The ID physician recommended a switch from intravenous to oral clindamycin for another 7 days, along with change from IV penicillin to intravenous ceftriaxone (for ease of outpatient administration), before completing treatment with oral amoxicillin. He was to have a total of 4 weeks of antimicrobial therapy.

Six weeks after discharge from hospital, most of the laboratory parameters were back to normal. On 8/11/10, the chemistries showed a sodium value of 137, potassium 3.8, glucose 106, BUN/creatinine 12/1.4, respectively. The liver tests showed normal values for SGOT 20, SGPT 21, alkaline phosphatase 78, and total protein and albumin of 7.3 and 4.1 mg/dL, respectively.

The DIC was resolved, and the echocardiogram showed improvement in the left ventricular EF to 40–45%, 3 months after admission, in spite of the patient not taking the prescribed low dose of carvedilol and lisinopril.

A number of points are worth noting in this case. This patient's problem began with a sore throat for which he received treatment at an urgent care center. One of his cousins, with whom he was in close contact, had a documented sore throat a few days before he became ill with sore throat. It is likely that he acquired the infection from him, although this cannot be proved definitively. It is also interesting to note that the organisms isolated from the blood and peritoneal abscess had similar susceptibility patterns, with both showing in vitro resistance of the GAS to azithromycin, the drug received by the patient for his sore throat. GAS has been known sometimes to be resistant to macrolide antibiotics [1]. The sensitivity pattern is shown in Box 11.2a.

Box 11.2a Antimicrobial susceptibility report[*]

Culture, body fluid: collected: 6/9/10
Specimen description: abdominal fluid swabs
Special requests: culture with sensitivity
   Result: 1+ growth: *Streptococcus pyogenes* (beta strep group A)
   Report status: final 06/12/10
Susceptibility
Organism: 1+ growth: *Streptococcus pyogenes* (beta strep group A)
Method: MIC
Ampicillin < =0.06 blood S
Azithromycin >2 blood R
Chloromycetin >16 blood R
Rocephin < =0.25 blood S
Cleocin < =0.06 blood S
Claforan < =0.25 blood S
Cefepime < =0.25 blood S
Zinacef < =0.25
Erythromycin >0.5 blood R
Levaquin 0.5 blood S
Penicillin G < =0.03 blood S
Tetracycline >4 blood R
Vancomycin 0.5 blood S

*The susceptibility report for the peritoneal fluid was similar to that of the blood cultures of the same day on 6/9/10 (2 of 4 BCs were positive for GAS). Please note that the susceptibility report is presented as reported by the laboratory. The listed trade names do not imply any endorsement.

The risk factors that led to this severe GAS infection in this patient are not clear, except of course the antecedent pharyngitis. The organism was not serotyped and so we cannot tell which strain or M serotype may have been responsible. Some predisposition may have been present even if we cannot pinpoint it.

Stevens has summarized the risk factors for development of severe GAS infections: minor trauma, injuries resulting in hematoma, bruising, or muscle strain; surgical procedures (e.g. suction lipectomy, hysterectomy, vaginal delivery, bunionectomy, bone pinning, breast reconstruction, cesarean section); viral infections (e.g. varicella, influenza), and use of non-steroidal anti-inflammatory drugs (NSAIDs) [2].

Our patient does not fit any of the above categories, with perhaps the remote possibility of minor trauma or muscle strain which could have been missed. He had worked very hard outdoors in the preceding 2 weeks in the heat. The sore throat (pharyngitis) appears to be the most likely source of the GAS infection. This patient had prompt and aggressive surgical intervention to address the peritonitis. This intervention was clearly life-saving.

- *Final diagnosis: Group A streptococcal toxic shock syndrome with multiorgan dysfunction.*

The clinical characteristics exhibited by this patient fit the definition in the literature for TSS (Box 11.2b, Table 11.2a) [2–4].

Box 11.2b Clinical guideline for diagnosis of GAS toxic shock syndrome [2–4]

Isolation of **GAS** from a normally **sterile site** (e.g. blood, cerebrospinal, pleural, or peritoneal fluid, tissue biopsy, or surgical wound) **plus**
Hypotension (systolic blood pressure ≤90 mmHg in adults or <5th percentile for age in children)
plus two or more of the following:
Renal impairment (creatinine in adults, ≥2 mg/dL; in children, two times upper limit of normal for age; in patients with pre-existing renal disease ≥twofold elevation over baseline)
Coagulopathy (e.g. thrombocytopenia, disseminated intravascular coagulation)
Liver involvement (e.g. ≥two times upper limit of normal for age of transaminases or bilirubin; in patients with pre-existing liver disease ≥twofold elevation over baseline)
Adult **respiratory distress** syndrome
Erythematous macular rash, may desquamate; and
Soft tissue necrosis (e.g. necrotizing fasciitis, myositis, or gangrene)

The case fatality rate for GAS TSS is higher than that for staphylococcal TSS.

The epidemiology and pathogenesis of GAS TSS have been summarized below [2,4].

- Severe invasive GAS infections are defined as bacteremia, pneumonia, necrotizing fasciitis, gangrenous myositis, or any other infection associated with the isolation of GAS from a normally sterile body site.
- GAS TSS is defined as any GAS infection associated with the acute onset of shock and organ failure.
- GAS TSS is mediated by toxins that activate the immune system, resulting in the release of large quantities of inflammatory cytokines that cause capillary leak and tissue damage, leading to shock and multiorgan failure.

Table 11.2a Clinical characteristics of toxic shock-like syndrome caused by group A streptococcus [2,4].

Characteristic	Comment
Age group affected	All ages; predominantly adults
Sex most often affected	No predilection
Portal of entry	Variable, generally skin or soft tissue (also vagina and respiratory tract)
Clinical features	Fever; hypotension; altered mental status, coma; signs of soft tissue infection that usually progresses to necrotizing fasciitis or myositis; active varicella with secondary infection; multiorgan dysfunction, often with acute respiratory distress syndrome
Bacteremia	Often present (about 60%)
Causative organism	Group A streptococcus
M serotype	Variable; 1, 3, 12, and 28 most common
Type of pyrogenic exotoxin	A, B, C, SSA, MF
Case fatality rate	30–60%

- The prevalence of GAS infections increased in the 1980s and has remained at approximately 3.5 cases per 100,000 people. Up to one-third of these cases develop GAS TSS.
- Persons of all ages may be afflicted with GAS TSS and most are not immuno-suppressed.
- Shock at the time of admission or within 4–8 hours is present in virtually all patients with GAS TSS.

Stevens has emphasized the importance of clindamycin as a key component of TSS treatment, especially as penicillins are most active in the logarithmic phase of growth of the organism, and have no toxin-binding capability [5]. Other advantages over penicillin include longer postantibiotic effect of clindamycin, as well as not being affected by inoculum size (a problem for penicillins).

Recommendations for treatment of streptococcal TSS [5]

- Optimal management of a patient with GAS TSS includes management of the complications of sepsis, aggressive surgical debridement if a site of infection is identified, and antibiotics for the underlying infection.
- Prompt and aggressive exploration and debridement of suspected deep-seated GAS infections.
- For patients who present with the clinical manifestations of GAS TSS prior to the identification of GAS by culture: empiric therapy with broad-spectrum antibiotics which should include clindamycin.
- Once the diagnosis of GAS TSS is established, recommend therapy with clindamycin (900 mg IV every 8 hours) in addition to penicillin G (4 million units IV every 4 hours).
- The duration of antibiotic therapy for GAS TSS should be individualized.

- Stevens suggests therapy with intravenous immune globulin (1 g/kg day 1, followed by 0.5 g/kg on days 2 and 3) in patients with GAS TSS [5].

Lessons learned from this case

- Prompt intervention with surgery was essential in saving the life of this patient.
- Even with his youth (only 18 years), recovery from this illness back to his baseline would still take many months. At 3 months, his left ventricular ejection fraction was still only 40–45%.
- He clearly had too many consultants because there was no clarity about a unifying diagnosis, at least initially.
- He ended up with many more tests and procedures than were needed for optimal care of his illness (just because he was seen by so many consultants).
- Again, a good and detailed history and physical exam are essential to an accurate, prompt, and complete diagnosis. Timing of intervention may be essential for a favorable outcome.
- Azithromycin (a macrolide antibiotic) is not a reliable antimicrobial agent for streptococcal pharyngitis (it did not work in this patient); the GAS was resistant to the agent.
- Overall, the patient had appropriate antimicrobial therapy. He was just too ill to recover quickly.
- The infectious diseases consult, although able to provide a unifying diagnosis, was rather too late in the course to make a meaningful impact on the critical stage of his illness. In these sick patients, the earlier the consult is placed, the better.
- We are reminded here again that GAS TSS is a very serious disease, with multiorgan dysfunction and a high mortality.

References

1 Stevens DL, Bisno AL, Chambers HF, et al. IDSA practice guidelines for the diagnosis and management of skin and soft-tissue infections. Clinical Infectious Diseases 2005; 41: 1373–406.

2 Stevens DL. Epidemiology, clinical manifestations, and diagnosis of streptococcal toxic shock syndrome. Available at: www.uptodate.com/online/content/topic.do?topicKey=gram_pos/5335&selectedTitle=2~18&source=search_result, accessed March 3, 2016.

3 Bisno AL, Stevens DL. Streptococcus pyogenes. In: Mandell GL, Bennett JE, Dolin R (eds) *Principles and Practice of Infectious Diseases*, 7th edn. Philadelphia: Churchill Livingstone, 2010, pp. 2593–610.

4 CDC. Case definitions for infectious conditions under public health surveillance. Morbidity and Mortality Weekly Report 1997; 46(RR-10): 38–40.

5 Stevens DL. Treatment of streptococcal toxic shock syndrome. Available at: www.uptodate
.com/online/content/topic.do?topicKey=gram_pos/7261&selectedTitle=27%7E150&
source=search_result, accessed March 3, 2016.

Case 11.3   A 45-year-old woman with rash, hepatitis, and lymphadenopathy

A 45-year-old black female was admitted through the ER in early May, 2000 with complaints of generalized weakness, rash, nausea, and vomiting. She had been ill with malaise for nearly a month, and rash had been present over the preceding 2 weeks. Because of the nature and severity of her illness, she was evaluated in consultation with the hematologist and gastroenterologist. With no clear or definitive diagnosis, infectious diseases consult was sought 12 days later on 5/18/00.

This patient had been ill with malaise for at least 1 month PTA. Fever was not initially apparent to the patient or the family. The rash had started in the upper arms and face 2 weeks PTA, before it spread to involve the rest of the body. There was itching associated with the rash. Two to 3 days PTA, with worsening weakness, nausea, and vomiting, she was brought to the ER for evaluation.

Epidemiologic history suggested no exposure to insects, camping, bites, stings, pets or any animals. There was no known change in detergents or use of cosmetics. No other family members were ill.

The review of systems was positive for rash with itching, vomiting, poor appetite, generalized malaise, weakness, and some sore throat.

The past medical history was significant for hypertension, arthritis, and seizure disorder since her teenage years.

She was married, was a stay-at-home mom, and had three daughters aged 11, 13, and 15 years. She was a non-smoker and non-drinker. She had allergy to penicillin (rash).

Medications PTA included phenytoin 100 mg three times a day, and losartan with hydrochlorothiazide daily.

On examination, she was drowsy but arousable, drifting off to sleep intermittently during the exam. She was weak but able to follow simple commands and answer simple direct questions.

Her vital signs were as follows: BP 96/55, RR 20, HR 80, temperature 97.3 °F (T_{max} the night before was 103.5 °F; 100.1 °F on admission). Height was 5'5", and she weighed about 160 pounds. Head and neck examination showed scalp dandruff and desquamation; oropharynx showed dry mucosa and carious teeth. The neck showed a 2 cm prominent palpable lymph node in the right posterior cervical area, plus other tiny neck nodes. The heart, lung, and abdominal exams were unremarkable. Lower extremities showed non-pitting edema. Lymph nodes were palpably enlarged in both axilla, right posterior cervical, and in both groins. They were moveable and non-tender. The neurologic exam showed a drowsy patient (on sedating meds – diphenhydramine), who was easily arousable. She had no focal findings, but diminished deep tendon reflexes.

The skin showed a diffuse, erythematous, non-blanching rash, involving the upper and lower extremities, especially forearms and feet, trunk, face, and chest, with some desquamation. There was no cellulitis noted. The thighs, shin, and foot showed diffuse erythroderma and desquamation.

- At this stage what differential diagnoses are you thinking of

Laboratory and other investigations

Blood cultures on 5/6/00 and 5/7/00 (4 out of 4) were negative. A culture from left ear drainage on 5/7/00 turned out to be positive for methicillin-sensitive *Staph. aureus*. Urinalysis on admission was unremarkable except for positive bile; stool occult blood test ×3 was negative. The CBC on admission showed a WBC of 8.9 and on 5/16/00, WBC count was 13.3; H/H 9.5 and 28.3, respectively; platelet count 371. Differential count showed 67% polymorphonuclear leukocytes (PMNs), 4% band forms, and 24% lymphocytes. There was no eosinophilia noted on the peripheral blood smear on any CBC done during the May, 2000 admission. The initial BUN and creatinine were 25 and 1.8, respectively. The serial liver panel tests are shown in Table 11.3a.

Chest x-ray on admission on 5/6/00 did not show infiltrates, but a follow-up x-ray on 5/14/00 suggested diffuse interstitial infiltrates. Ultrasound (US) evaluation of the lower extremities showed no evidence for deep venous thrombosis. Gallbladder US and upper GI series on 5/8 and 5/9/00 were reported as normal.

Bone marrow biopsy done on 5/16/00 showed normal cellular marrow with trilineage hematopoiesis and mildly decreased iron stores. No malignancy was identified.

Needle biopsy of the liver done on 5/19/00 showed increased lymphocytes in the periportal areas. There was steatosis present and foci of intralobular collections of lymphocytes. There was no increase in fibrosis. It was interpreted as mild chronic hepatitis.

Table 11.3a Serial liver panel recorded in May, 2000.

Liver test	5/11/00	5/17/00	5/19/00	5/21/00	5/24/00	Units	Reference range
SGOT (AST)	51	196	214	148	206	IU/L	15–46
SGPT (ALT)	97	132	123	110	141	IU/L	11–66
Alkaline phosphatase	791	1075	953	1226	1700	IU/L	38–126
Total bilirubin	0.56	2.01	5.15	6.57	3.32	mg/DL	0.2–1.3
Conjugated bilirubin	0.12	0.93	ND	ND	1.68	mg/DL	0.00–0.03
Albumin	1.7	2.0	1.8	1.7	2.1	g/DL	3.1–5.1
Total protein	4.8	6.6	6.3	5.9	7.1	g/DL	6.3–8.2

ALT, alanine aminotransferase; AST, aspartate aminotransferase; ND, not done or recorded; SGOT, serum glutamic oxaloacetic transaminase; SGPT, serum glutamic pyruvic transaminase.

Epstein–Barr virus (EBV) antibodies on 5/13/00 did not show elevation of IgM antibodies but serology for cytomegalovirus (CMV) IgM by immunofluorescent antibody test was found to be elevated at 1:80 units (N <1:10). No CMV IgG result was available.

Acute hepatitis serology on 5/8/00 for A, B, and C was negative. The prothrombin time (PT) on 5/16/00 was elevated at 17.5 seconds, reflecting some coagulopathy. HIV serology screen on 5/16/00 was negative while CD4 (helper/inducer T cell) was elevated at 68% or 2205 counts/cu mm (N 720–1440).

Quantitative immunoglobulin levels (IgA, IgG, and IgM) were moderately elevated, with IgE levels particularly high at 11,600 KU/L (N <160.0 KU/L). Antinuclear antibody (ANA) was mildly elevated at 1:80 (N <1:40). Qualitative rapid plasma reagin (RPR) test for syphilis was reactive at a low titer of 1:2, and C-reactive protein level was 5.1 mg/dL (N <0.8).

- *Do you now have a specific diagnosis or how have you broadened your differential diagnoses?*

Hospital course

A number of differential diagnoses were considered by the consultant: drug hypersensitivity reaction to phenytoin; autoimmune disease; reactive CMV disease; lymphoma or pseudo-lymphoma. Secondary syphilis was considered less likely because of the very low titer of RPR; typically with secondary syphilis, the RPR titer is high. The confirmatory fluorescent trepondemal antibody (FTA) test would turn out later to be negative. Liver or lymph node biopsy was to be recommended should the bone marrow (BM) biopsy done 2 days earlier turn out to be unrevealing or inconclusive. These procedures would only be done after correction of the coagulopathy.

Meanwhile, two specific recommendations were made to the patient care team.

- *Can you guess what these specific recommendations were?*

The patient was started on one new medication, while another was withdrawn. The fever abated but the liver enzyme levels did not immediately come down to normal. The transaminases and the alkaline phosphatase levels remained elevated, as noted in Table 11.3a. The graphs of the temperature recordings in the week before and after 5/18/00 are shown in Fig. 11.3a.

The patient improved enough clinically to be discharged home, afebrile, on 5/26/00.

The patient was hospitalized for 20 days, from 5/6/00 to 5/26/00.

Readmission of patient to the hospital

This patient was readmitted to the hospital 1 month later on 6/28/00 with recurrence of fever and weakness. One day before admission, she had been seen in the ER because of dysuria and frequency. Urinalysis had been found abnormal, with positive leukocyte esterase (2+), bacteria (2+), and 30 WBCs/high powered field

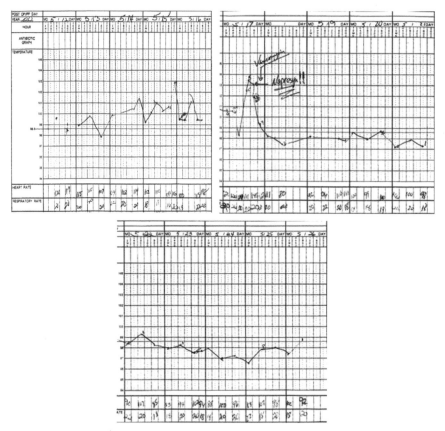

Figure 11.3a Temperature and vital signs recorded between 5/12/00 and 5/26/00.

(hpf). She was prescribed ciprofloxacin, along with promethazine for nausea and vomiting. A urine culture was not available.

On return to the ER the next day, on 6/28/2000, her temperature was 103.2 °F, and she was extremely weak and wobbly on her legs. She was admitted to the hospital for further work-up.

Hospital course

Recall that this patient had been earlier admitted for fever, rash, hepatitis, and lymphadenitis from 5/6 to 5/26/00. Although drug hypersensitivity was suspected, there had been no conclusive diagnosis at the time of discharge from the hospital.

Treatment was started for presumptive urinary tract infection (cystitis or pyelonephritis), based on the urinalysis (since no culture was available), with levofloxacin. The temperature came down promptly to between 99.0 °F and 100 °F.

Surgery was consulted to do a lymph node biopsy, to complete the work-up initiated during the May, 2000 admission, since the BM and liver biopsies were non-diagnostic. A biopsy of the right axillary lymph node was done on 7/3/00.

The temperature, which was 99.1 °F on that day, rose to 103.9 °F 2 days later on 7/5/00, in association with chills and tenderness of the armpit.

The infectious diseases consultant was asked to see the patient for this new-onset or recurrent fever.

Recent past history and review of systems showed that the recently elevated liver enzyme levels noted in May had come down to near normal, and that the renal insufficiency had resolved. The diffuse, reddish, peeling erythema (skin rash) was markedly improved, with left-over dry scales on the skin and scalp only. She had now only mild itching compared to the last admission in May.

The new problem was the urinary frequency and burning, and the fever that started just prior to this admission and recurred on 7/4/00.

On examination, she was alert and oriented, and moderately acutely ill. Her vital signs showed the following: BP 137/72, RR 22, HR 135, and temperature 103.9 °F; height 5'5", and weight 161.8 pounds. Head and neck exam showed poor oral hygiene. Cervical adenopathy was much improved, meaning smaller palpable nodes. Heart, lung, and abdominal exams were unremarkable. Upper extremities were unremarkable, while lower extremities showed minimal athlete's foot. Neurologic exam was grossly normal, with no focal deficits.

The skin showed desquamating scales on the scalp and body torso, but less prominent than in May. There was, however, no redness of the skin this time. The biopsy site in the right axilla was swollen and tender to touch; pinkish-brownish serous fluid was seen to ooze from the site as soon as the Steri-Strip dressing was removed.

Laboratory and other investigations

Computed tomography scan of the chest done on 6/29/00 showed mediastinal adenopathy and patchy infiltrates of the lung, which were considered chronic, based on a comparison with previous chest x-rays going back many years. CT scan of the abdomen confirmed a 3 cm mass in the pelvis consistent with uterine fibroids.

Complete blood count on 6/27/00 showed a WBC count of 9.6; on 6/28/00 WBC count was 10.1. Abnormal urinalysis of 6/27/00 was previously noted, without a culture. Erythrocyte sedimentation rate (ESR) on 6/29/00 was 45 mm/hour; HIV serology was negative. Serum IgE level on 6/29/00 was down to 1921.0 KU/L, still elevated but lower than the 11,600 noted on 5/15/00 (N < 160.0 KU/L). A single blood culture obtained on admission was negative. The gram stain obtained from the right axilla drainage on 7/5/00 showed moderate amounts of WBCs and gram-positive cocci in clusters and pairs. The culture from the lymph node biopsy on 7/3/00 was positive for 1+ methicillin-resistant *Staph. aureus* (MRSA). The CBC on 7/5/00 showed a WBC count of 14.9, H/H of 9.9 and 29.4, respectively, and platelet count of 320. Differentials: 80% PMNs, 12.7% lymphocytes, and 7% monocytes. BUN and creatinine were 6 and 1.2, respectively. Liver tests are noted in Table 11.3b.

The right axilla lymph node biopsy of 7/3/00 was described as "dermato-pathic lymphadenopathy" (a distinct form of reactive lymphoid hyperplasia seen

Table 11.3b Serial liver function tests measured after the May, 2000 admission.

Liver test	6/5/00	7/5/00	Units	Reference range
SGOT (AST)	33	17	IU/L	15–46
SGPT (ALT)	31	27	IU/L	11–66
Alkaline phosphatase	525	246	IU/L	38–126
Total bilirubin	1.44	NA	mg/DL	0.2–1.3
Conjugated bilirubin	0.00	NA	Mg/DL	0.00–0.03
Albumin	3.2	2.5	g/DL	3.1–5.1
Total protein	8.2	NA	g/DL	6.3–8.2

ALT, alanine aminotransferase; AST, aspartate aminotransferase; NA, not available; SGOT, serum glutamic oxaloacetic transaminase; SGPT, serum glutamic pyruvic transaminase.

in patients with benign chronic dermatoses and mycosis fungoides) [1]. Flow cytometry and other special studies, including chromosomal and tumor marker analyses done at a reference laboratory on the biopsied node, were, however, negative for malignancy.

The second admission was from 6/28/00 to 7/12/00.

Case discussion

The initial admission was from 5/6 to 5/26/00. The patient had fever, severe erythematous, desquamating, and pruritic rash, as well as hepatitis and significant lymphadenopathy in the neck, axilla, and inguinal areas. Extensive work-up, including BM and liver biopsies, did not reveal a definitive diagnosis at the time of discharge. The patient was, however, clinically improved at the time of discharge on 5/26/00, in spite of the persistently elevated liver enzymes. The fever was resolved.

A presumptive clinical diagnosis of "drug hypersensitivity syndrome" was made on 5/18/00 and a decision was made to do the following: stop phenytoin and begin systemic steroid (prednisone) therapy. Although the patient was also on antibiotics (levofloxacin and vancomycin) at the time, it is believed that the fever resolved because of the withdrawal of the anticonvulsant drug, phenytoin, and the use of systemic steroids.

This patient continued to improve after discharge on 5/26/00. It is not clear how long she stayed on the systemic steroids after the 5/26/00 discharge.

The readmission on 6/28/00, however, was because of acute urinary tract infection. It is notable that as early as 6/5/00, 3 weeks before readmission, the transaminases were back to normal, with a marked reduction in the level of the alkaline phosphatase (see Table 11.3b for liver tests on 6/5 and 7/5/00). The improvement continued through the second admission in June–July, 2000. The second spike of fever after the right axilla node biopsy was a complication of the procedure (Fig. 11.3b). MRSA infection of the wound/biopsy site was

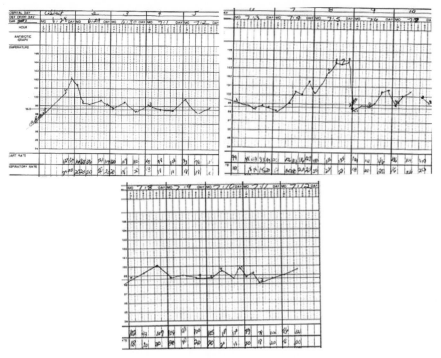

Figure 11.3b Fever pattern noted during the second admission: 6/28–7/12/00.

supported by drainage of the site after the biopsy, and the positive wound culture. The fever resolved with local wound care and antibiotic therapy.

After discharge from the hospital on 7/12/00, she was prescribed a combination of ciprofloxacin and trimethoprim/sulfamethoxazole by her primary care physician. At follow-up on 8/1/00, it was noted that she had developed a pruritic reaction; she finally completed her therapy for the axilla wound infection with clindamycin. It is not clear which of the antibiotics was responsible for the itching.

- *Final diagnosis: DRESS syndrome (drug rash with eosinophilia and systemic symptoms).*

DRESS syndrome: discussion and postscript

DRESS syndrome is a term introduced by Bocquet et al. in 1996 [2] and stands for drug reaction (or rash), with eosinophilia and systemic symptoms. Bocquet used this term to separate DRESS from other similar adverse drug reactions [2]. Typically, patients have fever, rash, lymphadenopathy, and internal organ involvement with systemic manifestations. The most commonly involved organs are the liver, followed by the kidney and lungs [3]. The syndrome manifests as a severe idiosyncratic reaction that is characterized by a long latency (weeks or longer) after exposure to the offending drug.

Other synonyms for DRESS include drug hypersensitivity syndrome (DHS), anticonvulsant hypersensitivity syndrome (AHS), drug-induced pseudo-lymphoma, and drug-induced delayed multiorgan hypersensitivity syndrome (DIDMOHS) [2,4].

A review by Cacoub et al. analyzed the clinical course and treatment of a large number of cases and noted a total of 44 drugs that were associated with DRESS over a 12-year period [5]. The aromatic anticonvulsants (phenytoin, phenobarbital, and carbamazepine) and sulfonamides are among the most common causes of DRESS syndrome [6]. Mortality in DRESS has been reported to range between 5.2% and 10% [5,6].

Our patient met more than the required three European Registry of Severe Cutaneous Adverse Reactions (RegiSCAR) criteria for DRESS [7], but did not have eosinophilia. She was hospitalized, had a rash suspected to be due to drug hypersensitivity, had fever, enlarged lymph nodes, and internal organ involvement with hepatitis and renal injury.

Although eosinophilia is common, it may not be seen in up to 50% of patients with DRESS syndrome [4]. Our patient responded to withdrawal of phenytoin and treatment with systemic steroid. The BM, liver, and lymph node biopsies did not show any malignancy or other specific diagnosis. The hepatitis resolved, and the fever and rash also resolved. It is interesting that she developed pruritus after she received a combination of ciprofloxacin and trimethoprim/sulfamethoxazole. It is speculative to consider that the sulfonamide may have been responsible for the pruritus.

Although this patient has had a seizure disorder since her teenage years, it was not clear from the history which other anticonvulsants she had been on in the past. However, she claimed to have been on phenytoin for at least several years before the admission in May, 2000. Whether the use of other medications, e.g. sulfonamide, precipitated her phenytoin-induced hepatitis and hypersensitivity cannot be proven with certainty. The concomitant use of drugs like sulfonamides has been shown to affect the clearance of aromatic anticonvulsants from the liver, and prolong the effect of anticonvulsants like phenytoin [8].

The latency period from initiation of phenytoin and DRESS in this instance would seem to be longer than typically reported. Her symptoms, however, resolved after discontinuation of phenytoin, and treatment with steroids. The elevated liver enzymes, lymphadenopathy, rash, and fever also resolved.

She was counseled to consider herself allergic or hypersensitive to phenytoin. She was to be observed off all anticonvulsants for a while, and be re-evaluated by a neurologist for an alternative treatment should she develop a new seizure. Since cross-reactivity among the major anticonvulsants is common [9], we thought that the decision regarding alternative therapy should be left to the specialist.

Lessons learned from this case

- The latency period between initial drug exposure and symptoms of DRESS can be longer than the usual weeks to months. In this patient, symptoms developed after years of drug exposure, and not months.
- A complete history of all medications ingested (past, recent, and current), including over-the-counter drugs, is important in any attempt to address adverse drug reactions.
- Delay in diagnosis of DRESS is common, because of the need to rule out other serious potential differential diagnoses: lymphoma, other malignancies, autoimmune disease, secondary syphilis, drug hypersensitivity, or other serious systemic infections.
- As in this case, the work-up can be expensive and prolonged.
- An early decision to withdraw a non-critical drug may have averted the need for multiple invasive biopsies, as was done in this patient. This patient had biopsies of the BM, lymph node, and liver, all of which were non-diagnostic.

References

1 Burke JS, Colby TV. Dermatopathic lymphadenopathy: comparison of cases associated and unassociated with mycosis fungoides. American Journal of Surgical Pathology 1981; 5: 343–52.

2 Bocquet H, Bagot M, Roujeau JC. Drug-induced pseudolymphoma and drug hypersensitivity syndrome (Drug Rash with Eosinophilia and Systemic Symptoms: DRESS). Seminars in Cutaneous Medicine and Surgery 1996; 15(4): 250–7.

3 Joo HL, Hye-Kyung P, Jeong H, et al. Drug rash with eosinophilia and systemic symptoms (DRESS) syndrome induced by celecoxib and anti-tuberculosis drugs. Journal of Korean Medical Science 2008; 23(3): 521–5.

4 Walsh SA, Creamer D. Drug reaction with eosinophilia and systemic symptoms (DRESS): a clinical update and review of current thinking. Clinical and Experimental Dermatology 2011; 36(1): 6–11.

5 Cacoub P, Musette P, Descamps V, et al. The DRESS syndrome: a literature review. American Journal of Medicine 2011; 124 (7): 588–97.

6 Tas S, Simonart T. Management of drug rash with eosinophilia and systemic symptoms (DRESS): an update. Dermatology 2003; 206: 353–6.

7 European Registry of Severe Cutaneous Adverse Reactions to Drugs and Collection of Biological Samples (RegiSCAR). Available at: http://www.regiscar.org/, accessed March 3, 2016.

8 Hansen JM, Kampmann JP, Siersbæk-Nielsen K, et al. The effect of different sulfonamides on phenytoin metabolism in man. Acta Medica Scandinavica 1979; 205(S624): 106–10. Available at: http://onlinelibrary.wiley.com/doi/10.1111/j.0954-6820.1979.tb00729.x/abstract, accessed March 3, 2016.

9 Shear NH, Spielberg SP. Anticonvulsant hypersensitivity syndrome. In vitro assessment of risk. Journal of Clinical Investigation 1988; 82(6): 1826–32. Available at: www.ncbi.nlm.nih.gov/pmc/articles/PMC442760/, accessed March 3, 2016.

CHAPTER 12

Skin and Soft Tissue Infections Seen Post Hurricane Katrina in 2005

Introduction

During Hurricane Katrina that hit the United States Gulf Coast in the early morning of August 29, 2005, our hospital provided shelter for patients who could not be discharged, and some personnel stayed back to provide medical services for these patients. After the hurricane, we received injured patients from many surrounding towns, shelters, and damaged homes. Our hospital stayed open to provide care for injured patients who were transferred from other sites, or came on their own to the hospital. Several of these patients had soft tissue infections, illustrated in several case presentations below.

Case 12.1   A 76-year-old Caucasian female with leg laceration

A 76-year-old Caucasian female was admitted on 9/1/05, 3 days after Hurricane Katrina. Four days earlier on 8/28/05, she had fallen at home and bruised or lacerated her left shin and foot. Several hours after the hurricane hit on 8/29/05, she had gone around the house to clean up, wading into water in the process. The next day (8/30/05), she came to the emergency room (ER) where the laceration in her left shin was stitched and she was given cephalexin, and hydrocodone for pain. She complained of leg pain and difficulty getting around. She took a single dose of the cephalexin, but lost or misplaced her pain medication (hydrocodone). Over the next 24 hours, she became increasingly tired and sleepy, and a fever was noted.

She was brought back to the hospital ER on 8/31/05 at 2043 hours for further assessment. Following initial evaluation and some x-rays, she was admitted in the early hours of 9/1/05 to the medical intensive care unit (MICU).

Cases in Clinical Infectious Disease Practice: Obtaining a Good History from the Patient Remains the Cornerstone of an Accurate Clinical Diagnosis: Lessons Learned in Many Years of Clinical Practice,
First Edition. Okechukwu Ekenna.
© 2016 John Wiley & Sons, Inc. Published 2016 by John Wiley & Sons, Inc.

The past medical history and underlying diseases included the following: known Alzheimer's disease for 3 years; osteoarthritis of the knees and hands; hysterectomy many years previously; and colonoscopy several months earlier. She was a poor historian because of poor memory (Alzheimer's) and there was no reliable information on the status of her tetanus immunization.

Family and social history was significant for the following: she was married and lived with her very attentive and loving husband. She did not drink or smoke. She was allergic to penicillin (whelps and rash). She had been prescribed cephalexin 2 days before admission, but took only one or two doses of the drug. Her other medications included hydrocodone, memantine, and donepezil.

The physical examination showed her to be alert to person and able to follow simple commands, although she had generally poor memory. She was in no acute distress. Her vital signs showed the following: blood pressure (BP) 108/53, respiratory rate (RR) 24, heart rate (HR) 100, and temperature 101.9 °F; height 5′2″, weight 129.4 pounds. The head, neck, heart, lung, and abdominal exams were unremarkable. The leg examination showed sutured laceration of the left mid shin, while the ankle and foot showed areas of ecchymotic changes and a large blister on the left foot dorsum and medial ankle. Peripheral pulses were palpable, but pitting edema was noted in the leg (Fig. 12.1a).

Laboratory and radiologic findings

Chest x-ray showed no pneumonia while the x-ray of the left foot and ankle showed no fractures. Gram stain done on the aspirate of the leg blister showed no organisms and no white blood cells (WBCs). Chemistries revealed a sodium level of 139, potassium 4.1, glucose 155 mg/dL, and blood urea nitrogen (BUN)/creatinine of 37 and 1.8, respectively (an estimated glomerular filtration rate of 29 mL/minute). The complete blood count (CBC) showed a WBC count of 12.7, hemoglobin and hematocrit (H/H) of 15.6 and 47.3, respectively, mean corpuscular volume (MCV) of 99, and platelet count of 207,000. The differential count showed 67% polymorphs, 16% band forms, 5% atypical lymphocytes, and 12% monocytes. Urinalysis showed a specific gravity of 1.020; 3+ blood, >30 red blood cells (RBCs), and 3–5 WBCs. Blood cultures were negative. The culture of the aspirated leg blister was reported positive in 24 hours, and confirmed in 2 days.

• *What is the likely diagnosis? What organism do you think was found in the culture?*

Follow-up care

Infectious diseases consult was placed the day after admission, and a recommendation was made for surgical debridement of the leg. The patient was taken to surgery for debridement, and required several debridement procedures over the

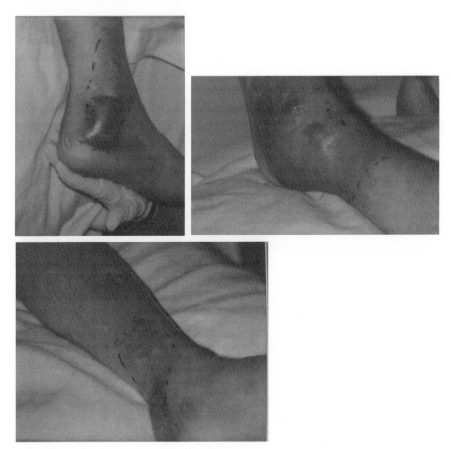

Figure 12.1a Hemorrhagic blisters noted at the medial and lateral left foot, ankle and heel. Photos taken on 9/2/05 (reproduced with permission). (*See insert for color information*)

next 2 weeks. She eventually had skin grafting performed on 9/22/05, 3 weeks after admission to the hospital. She was finally discharged on 10/11/05, 40 days after initial admission. Her care included antibiotics (intravenous ceftazidime), and oral doxycycline and levofloxacin. Her prolonged hospital admission included placement (stay) at the rehab center for better care of this elderly lady with Alzheimer's dementia, as well as for hyperbaric oxygen therapy. A photograph of the left leg 4 days after the initial surgical debridement on 9/2/05 is shown in Fig. 12.1b.

• *Have you now changed your diagnosis?*

The working diagnosis was necrotizing fasciitis of the left leg and foot. The microbiologic diagnosis and the organism isolated will be discussed later.

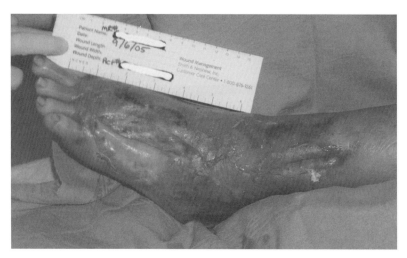

Figure 12.1b Debridement of left leg: one of several done on this patient. Photo taken on 9/6/05, 4 days after the initial debridement (reproduced with permission).

Case 12.2   A 76-year-old man with rapid-onset cellulitis

A 76-year-old Caucasian male was admitted on 8/31/05, 2 days after Hurricane Katrina. He had been trapped in his house (unable to get out) during the storm, with water rising up to his neck. It took several hours for the water to recede enough for him to be able to get help. By the next day (8/30/05), he noted redness of his left leg, associated with leg pain. He also felt feverish and had chills. Over the next 24 hours, there was redness of his leg, with blister formation. He was seen in the ER of the hospital and admitted on 8/31/05.

Underlying diseases included osteoarthritis, hypertension, irregular heartbeat, macular degeneration, diminished hearing, and cholecystectomy years ago. The patient was overweight. He was a former heavy smoker of two packs a day for 30 years, but had not smoked in 20 years. He had no known drug allergies.

He was managed initially with various antimicrobial agents that included ceftazidime, doxycycline, and vancomycin (to cover for possible methicillin-resistant *Staph. aureus* [MRSA]). Blood cultures were obtained on admission on 8/31/05.

Seven days after admission, infectious diseases consult was placed to address the unresolved cellulitis, as well as a positive blood culture result.

On examination on 9/7/05, the patient was alert and oriented. His vital signs were as follows: BP 139/71, RR 12, HR 93, and temperature 97.6 °F; height 6'0", and weight 270 pounds. He was an obese, pleasant gentleman in no acute distress. The main abnormality noted on the physical examination was cellulitis of

Figure 12.2a Edema and blisters noted in the left foot and ankle. Photo taken on 9/7/05, 7 days after admission (reproduced with permission).

the left lower extremity. Photographs of the left leg taken on 9/7/05 are shown in Fig. 12.2a.

- *What is the likely etiologic diagnosis?*
- *What was the recommendation of the infectious diseases consultant?*

Laboratory and radiologic findings

The CBC on admission showed a WBC count of 11.4, H/H 12.3 and 36.4, respectively, MCV 88.0, and platelet count 164,000. The differential counts were as follows: 83% granulocytes, 5.8% lymphocytes, and 8.8% monocytes. Sedimentation rate was 10 mm/hour. Prothrombin time and partial thromboplastin times were normal. Chemistries showed a sodium level of 141, potassium 3.7, glucose 127 mg/dL, and BUN/creatinine of 30 and 2.0, respectively. Serum glutamic oxaloacetic transaminase (SGOT) was elevated at 167 IU/L (normal 15–41). A swab culture of the left leg (not blister) obtained on 9/1/05 was positive for methicillin-sensitive *Staph. aureus* (MSSA) while 2 of 2 blood cultures were reported positive for a gram-negative rod that was oxidase positive. The blood culture organism was pan sensitive to all the agents tested, including ampicillin.

- *Can you guess what organism was finally identified in the blood culture?*

Hospital course

The infectious diseases consultant recommended surgical debridement of the left leg blistered cellulitis. The patient required several debridement procedures of the leg, as well as hyperbaric oxygen therapy. He also received wound care, including wound vacuum-assisted closure (VAC) during the course of the prolonged treatment and healing. He was eventually discharged 3 weeks later, on 9/22/05, to outpatient follow-up. Follow-up labs on 9/7/05 subsequently showed normal chemistries, including normal liver and renal function tests. Figures 12.2b and 12.2c show photos of the left leg taken at various times after debridement. The patient was followed for more than 6 weeks in the outpatient setting.

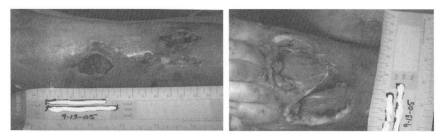

Figure 12.2b Necrotizing fasciitis left shin and foot, 6 days after initial debridement. Photo taken 9/13/05 (reproduced with permission). (*See insert for color information*)

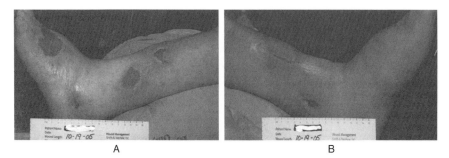

A   B

Figure 12.2c Slowly healing leg wounds 6 weeks after initial debridement of necrotizing fasciitis. Photo taken on 10/19/05. A. Lateral view. B. Medial view of left leg and foot (reproduced with permission).

Comments and final microbial diagnoses

The previous patient, the 76-year-old Caucasian female, had negative blood cultures but a positive wound culture for *Vibrio vulnificus*. The second patient in this category was a 76-year-old male with positive blood cultures for *Vibrio vulnificus*, but negative wound culture for *Vibrio* organism. The blister was not cultured initially until surgery on 9/7/05, 7 days after admission. He had been on antibiotics since admission. The MSSA cultured from the skin of the leg on 9/1/05 was clearly not the cause of the necrotizing fasciitis, but likely a skin colonizer. Both patients were elderly, and required several weeks to recover from their necrotizing fasciitis. Extended wound care and hyperbaric oxygen therapy were utilized in the management of these patients. Both recovered and at least one of these patients had some sort of skin grafting to cover the skin defects of the legs.

Case 12.3   A 61-year-old man with diabetic neuropathy

A 61-year-old Caucasian male was admitted on 8/31/05, 2 days after Hurricane Katrina, with lacerations on the feet and ankles.

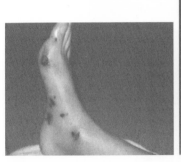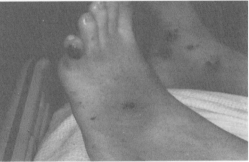

Figure 12.3a Multiple skin erosions caused by nail and other lacerations inside wet boots in a diabetic with peripheral neuropathy. Photo taken on 9/2/05, 2 days after admission.

Like many residents on the Gulf Coast, he had water surge into his house. He had worn rubber boots with wet feet, walking around and cleaning up after the storm. However, because of his severe diabetic peripheral neuropathy, he was unaware of nail(s) that had punctured his boots and stuck to his feet all day as he worked and walked around. Twenty-four hours later, when the boots were taken off, he noted multiple bruises and lacerations in both ankles and feet. He was persuaded to come to the hospital for further evaluation. He was admitted.

His examination was significant for skin abrasions, erosions, and lacerations, some of which looked like superficial decubitus ulcers (Fig. 12.3a).

Skin and blood cultures were negative. The patient was discharged home 6 days later, improved. The skin lesions looked like decubitus ulcers, abrasions, and erosions, and the clinical picture was not typical for *Vibrio* infection.

Case 12.4   A 45-year-old man from a refugee camp with calf laceration

A 45-year-old Caucasian male was admitted on 9/3/05 for a left calf infection following a laceration. He was a transfer from a refugee camp in southern Mississippi where he had been housed in a shelter following the hurricane that destroyed homes on the Gulf Coast on 8/29/05.

His initial injury occurred when he slipped coming down from the roof (where he had taken shelter) after the high water receded. He had been trapped on the roof for over 24 hours after the hurricane. He suffered scraping of his left posterior calf when he slipped on the way down from the roof. He was later transferred to a camp, where he shared space with others in a temporary shelter, described as an "unhygienic camp environment." There had been limited water supply, as well as electric power following the storm. He was transferred to our facility for further care because he developed a fever and leg pain.

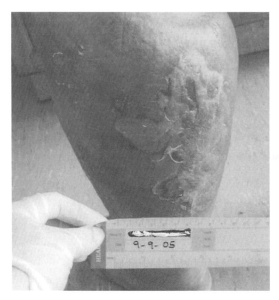

Figure 12.4a Posterior calf non-hemorrhagic blisters following leg laceration. Photo taken on 9/9/05.

On examination, he was a morbidly obese man, 450 pounds in weight, and with a height of only 5′4″. He had not had a bath in several days, nor had any change of clothing. He was sweaty and dirty. The temperature was 102.7 °F and his left leg showed non-hemorrhagic ruptured blisters with purulence and cellulitis. Figure 12.4a depicts redness, swelling, and blisters consistent with cellulitis and abscess.

Blood cultures were negative. The CBC showed a WBC count of 20,700/mm^3. Treatment was chosen to cover for presumptive staphylococcal and streptococcal infection. The patient was discharged 7 days later, in improved condition, back to his home town for continued care.

General comments on soft tissue infections in relation to Hurricane Katrina

A summary of the characteristics of soft tissue infections seen before and after Hurricane Katrina in 2005 is shown in Tables 12.1a and 12.1b, including outcome data.

Case discussion

During the period after the hurricane, there was some concern that unusual infectious diseases might turn up. In our hospital we saw many patients with

Table 12.1a Characteristics of serious soft tissue infections seen before Hurricane Katrina (July and August, 2005).

S/N	Admission dates (LOS)	Age/sex	Underlying disease/ risk factors	Blood culture (BC) results	Other site cultures	Outcome: died/survived (D/S)
1	7/20–8/10/05 22 days	45 C/M	Shrimper, leg abrasion, wade into water, pain & swelling; BP 83/41, WBC 34K; HCV/cirrhosis	2/2 pos. for *V. vulnificus* on 7/20/05	I&D L leg: 7/22 & 7/28 neg. cultures	S
2	8/15–8/25/05 10 days	66 C/M	Construction worker, injured R arm; cellulitis, rheumatoid arthritis, prednisone, Enbrel, COPD	3/3 pos. for *V. parahaemolyticus* on 8/15/05	I&D R arm 8/17/05 neg. culture	S Seen as outpatient only (9/28 & 10/5/05)
3	8/23–8/24/05 <1 day	36 C/F	HIV, HCV; salt water contact; alcoholic cirrhosis, drug addiction, leg pain & sores with blister, mottled skin, DIC, septic shock, BP 88/43	1/1 BC pos. for *V. vulnificus* on 8/23/05	L ankle pos. for *V. vulnificus*, *P. aeruginosa*, *P. mirabilis*, MSSA, GBS, *E. faecalis*	D Died within 24 hours of admission of fulminant sepsis

C, Caucasian; COPD, chronic obstructive pulmonary disease; F, female; GBS, group B *Streptococcus*; LOS, length of stay; M, male; MSSA, methicillin-sensitive *Staphylococcus aureus*; S/N, serial number of patients.

Table 12.1b Characteristics of serious soft tissue infections seen post Hurricane Katrina (after August 29, 2005).

S/N	Admission dates (LOS)	Age/sex	Underlying disease/risk factors	Blood culture (BC) results	Other site cultures	Outcome: died/survived (D/S)
4	8/30–8/31/05 1 day	82 B/M	Refused to evacuate during Katrina; found in the mud: near drowning, with aspiration pneumonia; in CHF, respiratory failure, and confused; CPK 1838, BNP 1130	2/2 pos. for V. vulnificus on 8/30/05	Not done or not available	Dead in 24 hours following admission
5	8/31–9/22/05 22 days	76 C/M	Water rose to neck in house; laceration to L leg; fever, chills, cellulitis with blister within 24 hours	2/2 pos. for V. vulnificus on 8/31/05	L foot and leg pos. for MSSA on 9/1/05	S Multiple I&Ds, HBO Rx, antibiotics
6	9/1–10/11/05 41 days	76 C/F	Fell at home 1 day before storm, lacerated leg; waded in contaminated water cleaning up after the storm	BCs of 9/1/05 negative	L foot/leg blister culture pos. for V. vulnificus on 9/1/05	S Prolonged hospital stay with multiple I&Ds, HBO, and antibiotics
7	8/31–9/6/05 6 days	61 C/M	Water surge in house; DM with severe peripheral neuropathy; wet boots all day, with nail puncture unnoticed	2/2 BCs drawn 8/31/05 negative	Culture: R & L foot and ankle skin erosions/abrasions negative, but presumed Staph	S
8	9/3–9/10/05 7 days	45 C/M	Morbidly obese, 450 lb, 5'4" refugee from a shelter; stranded on rooftop for >24 hours; lacerated calf on way down	BCs x 3 negative, WBC 20.7, temp 102.7°F	L leg/calf cellulitis No cultures done Presumptive Strep or Staph	S Airlifted from unhygienic camp to the hospital for care
9	9/4–9/9/05 5 days	17 C/M	Stepped on broken beer bottle in brackish water 8/27/05; I&D, then discharged and recalled for + culture	Negative BC of 8/27/05	R foot I&D 8/28 later pos. for V. fluvialis	S Called back for foot pain and pos. culture
10	9/4–9/12/05 8 days	79 C/M	On tree top for hours as water rose; spent days in shelter; arm, hand, body lacerations; cirrhosis/shunt surgery	2/2 B's of 9/4/05 negative	L arm/hand cult pos. for Alcaligenes	S Received ciprofloxacin before admission

B, black; BNP, basic natriuretic protein; C, Caucasian; CHF, congestive heart failure; COPD, chronic obstructive pulmonary disease; CPK, creatine phosphokinase; F, female; HBO Rx, hyperbaric oxygen treatment; I&D, incision and debridement; LOS, length of stay; M, male; MSSA, methicillin-sensitive *Staphylococcus aureus*; S/N, serial number of patients.

skin and soft tissue infections related to trauma (lacerations, cuts, and puncture wounds). These infections reflected increases in the number of usual infections typically present in our environment, and not exotic diseases.

The Centers for Disease Control and Prevention (CDC) has noted that widespread outbreaks of infectious diseases after hurricanes are not common in the United States [1]. Rare and deadly exotic diseases, such as cholera or typhoid, do not suddenly break out after hurricanes and floods in areas where such diseases do not naturally occur [1]. This prediction was borne out by the actual infectious diseases reported from the Gulf Coast after Hurricane Katrina (Table 12.1c) [2].

Communicable disease outbreaks of diarrhea and respiratory illness can occur when water and sewage systems are not working and personal hygiene is hard to maintain as a result of a disaster. In the report noted in Table 12.1c, dermatologic conditions and diarrheal diseases were commonly reported, including insect bites but not respiratory diseases. Because cholera and typhoid are not commonly found in the Gulf States area, it was very unlikely that they would occur after Hurricane Katrina. There was in fact no outbreak of these illnesses. Twenty-four cases of *Vibrio* illnesses were reported to the CDC, with six deaths [2]. A breakdown of the cases of *Vibrio* illnesses seen in Louisiana and Mississippi between 8/29 and 9/11/05 is illustrated in the MMWR publication of September 23, 2005, and in Fig. 12.5a [3].

Clinical and laboratory characteristics of *Vibrio* cellulitis and sepsis

Wound infection with *Vibrio* is typically evidenced by a rapid-onset cellulitis (1–2 days), following skin break or injury, leading to necrotizing fasciitis. Extreme pain at the involved limb and systemic illness, including shock, can occur. The skin may show peculiar blisters (often hemorrhagic) in association with the cellulitis, as was noted in patients 5 (case #12.2) and 6 (case #12.1) reported here (see Table 12.1b). Blood cultures are also rapidly positive (1–2 days) in seriously ill patients, while other site cultures are also positive, usually within 1–2 days. Rapid death is typical with overwhelming sepsis in those with severe underlying diseases (severe liver disease or other immunosuppression) (see patients 3 and 4 in Table 12.1a and 12.1b).

Serious infections with *Vibrio* organisms lead to prolonged hospitalization in those who survive, especially with *V. vulnificus* (patients 5 and 6 spent 22 and 41 days, respectively, in the hospital). Multiple surgical debridement, appropriate antibiotics, skin grafting, and, on occasion, HBO therapy may be required (as in patients 1, 5, and 6). Those with severe underlying disease arriving in septic shock (e.g. as in patients 3 and 4) have a rapid demise.

Successful treatment requires a combination of antibiotics, surgery, and sepsis/shock management that should be initiated early in the illness.

Table 12.1c Number of cases of selected diseases and conditions reported in evacuees and rescue workers during the 3 weeks immediately after Hurricane Katrina made landfall – multiple states, August–September 2005.

Disease/Condition*	No. of cases	States reporting	Population
Dermatologic conditions			
Infectious			
Methicillin-resistant *Staphylococcus aureus* infections	30 (3 confirmed)	Texas	Evacuees
Vibrio vulnificus and *V. parahaemolyticus* wound infections	24 (6 deaths)	Arkansas, Arizona, Georgia, Louisiana, Mississippi, Oklahoma, Texas	Evacuees
Tinea corporis	17	Mississippi	Rescue workers
Noninfectious			
Arthropod bites (likely mite)	97	Louisiana	Rescue workers
Diarrheal disease			
Acute gastroenteritis, some attributed to norovirus	Approximately 1,000	Louisiana, Mississippi, Tennessee, Texas	Evacuees
Nontoxigenic *V. cholerae* O1	6	Arizona, Georgia, Mississippi, Oklahoma, Tennessee	Evacuees
Nontyphoidal *Salmonella*	1	Mississippi	Evacuee
Respiratory disease			
Pertussis	1	Tennessee	Evacuee
Respiratory syncytial virus	1	Texas	Evacuee
Streptococcal pharyngitis	1	Texas	Evacuee
Tuberculosis	1	Pennsylvania	Evacuee
Other condition			
Presumed viral conjunctivitis	Approximately 200	Louisiana	Evacuee

*Other diseases and conditions, for which the number of cases was unknown, included scabies: circumferential lesions at waist; contact dermatitis; erythematous, papular, pustular rash consistent with folliculitis; immersion foot; prickly heat; influenza-like illness and upper respiratory infections; and head lice.

Source: CDC: www.cdc.gov/mmwr/preview/mmwrhtml/mm54d926a1.htm.

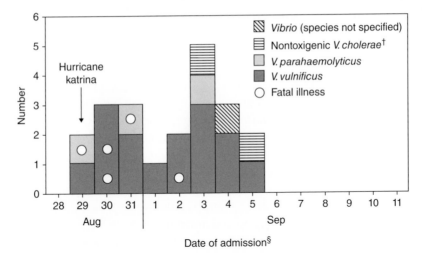

* N = 22; Alabama, a third state under surveillance, reported no cases.
† Nontoxigenic *V. cholerae* illnesses represent infections entirely distinct from the disease cholera, which is caused by toxigenic *V. cholerae* serogroup O1 or O139.
§ Date of admission was not available for one Louisiana resident. In cases that did not require hospitalization, the date represents the first contact with a health-care provider for the illness.

Figure 12.5a Cases of post-Hurricane Katrine *Vibrio* illness among residents of Louisiana and Mississippi, by date of hospital admission – United States, 8/29–9/11/05. Source: www.cdc .gov/mmwr/preview/mmwrhtml/mm5437a5.htm.

Vibrio infections peak in the warm summer months, as was the case when Hurricane Katrina occurred on 8/29/05.

Isolated *Vibrio* organisms seen in human disease in our hospital over the past 18 years have usually been found to be very sensitive to multiple antimicrobial agents. The third-generation cephalosporins, quinolones, tetracyclines, and other classes are effective clinically and in in vitro susceptibility tests. On occasion, ampicillin may be resistant. Patients die usually not because of antimicrobial resistance but for other reasons (overwhelming sepsis or immunosuppression). The most virulent *Vibrio*, *V. vulnificus*, has been found to be equally sensitive to the usually available antimicrobial agents.

Most *Vibrio* species have the following characteristics: they are gram negative, facultatively anaerobic, straight, curved, or comma-shaped rods that are catalase and oxidase positive [4]. The typical *Vibrio* organism shows a rapid growth (1–2 days) on blood agar plate, or special media, thiosulfate-citrate-bile salts-sucrose (TCBS), and is motile. Other related water-borne gram-negative organisms (*Aeromonas*, *Plesiomonas*, and Enterobacteriaceae) may be differentiated by their degree of sodium (salt) requirement for growth or stimulation, sensitivity to the vibriostatic compound (O/129), substrate fermentation, and growth on TCBS [4].

Addendum

A few cases of unusual or unexplained pruritic dermopathy, along with neuropsychiatric complaints, were also seen post Hurricane Katrina. The CDC set up a task force in 2006 to investigate the increased number of inquiries received regarding this unexplained cutaneous condition (crawling, biting, stinging, and other vague complaints). In addition to skin manifestations, some sufferers also reported fatigue, mental confusion, short-term memory loss, joint pain, and changes in vision. These symptoms were clinically not true soft tissue infections.

The results of the comprehensive CDC study of unexplained apparent dermopathy have been published [5,6]. These studies demonstrated no evidence for an infectious cause of this syndrome, and no evidence of an environmental link. Although the cause of this syndrome is unknown, a common viewpoint of medical professionals, including most dermatologists, is that the symptoms are most compatible with "delusional parasitosis." Another name associated with this disorder is "Morgellons."

Additional comments on disaster conditions (hurricane, unsanitary/refugee camp)

- The most common infectious diseases are those already present in the environment before the event.
- Such diseases may be magnified under those conditions (skin abscesses, scabies or mite infestations, diarrhea, and respiratory diseases).
- Exotic diseases are unusual, unless specifically introduced into the environment or camp (as happened in Haiti with cholera, following the earthquake of January, 2010) [7].
- Unsanitary conditions, overcrowding, and lack of a clean or adequate water supply facilitate transmission of many communicable or contagious diseases.

References

1 After a Hurricane: Key Facts About Infectious Disease. Available at: http://emergency.cdc .gov/disasters/hurricanes/pdf/infectiousdisease.pdf, accessed March 3, 2016.

2 Infectious Disease and Dermatologic Conditions in Evacuees and Rescue Workers after Hurricane Katrina – Multiple States, August–September, 2005. Morbidity and Mortality Weekly Report 2005; 54(Dispatch): 1–4. Available at: www.cdc.gov/mmwr/preview/ mmwrhtml/mm54d926a1.htm, accessed March 3, 2016.

3 Vibrio Illnesses After Hurricane Katrina – Multiple States, August–September 2005. Morbidity and Mortality Weekly Report 2005; 54(37); 928–31. Available at: www.cdc.gov/mmwr/ preview/mmwrhtml/mm5437a5.htm, accessed March 3, 2016.

4 Abbott SL, Janda M, Johnson JA, Farmer III, JJ. *Vibrio* and related organisms. In: Murray PR, Baron EJ, Jorgensen JH, Landry ML, Pfaller MA (eds) *Manual of Clinical Microbiology*, 9th edn. Washington, DC: ASM Press, 2007, pp. 723–33.

5 CDC.CDC Study of an Unexplained Dermopathy. Available at: www.cdc.gov/ unexplaineddermopathy/, accessed March 3, 2016.

6 Pearson ML, Selby JV, Katz KA, et al. Clinical, epidemiologic, histopathologic and molecular features of an unexplained dermopathy. PLoS ONE 2012; 7(1): e29908. Available at: www .plosone.org/article/info%3Adoi%2F10.1371%2Fjournal.pone.0029908, accessed March 3, 2016.

7 Chin CS, Sorenson J, Harris JB, et al. The origin of the Haitian cholera outbreak strain. New England Journal of Medicine 2011; 364(1): 33–42.

CHAPTER 13

Other Miscellaneous Infections

Case 13.1   A 17-month-old child with facial swelling and adenopathy

A 17-month-old Caucasian female was admitted on 5/13/09 with a 2-month history of chronic and recurrent right-sided facial swelling and lymphadenopathy.

The history of neck swelling started 2 months earlier when the patient was traveling with her parents in Texas. She woke up on 3/8/09 with a swelling involving the right side of the face. At a visit to a local emergency room, they were told it was "staph" infection, and cephalexin and trimethoprim/sulfamethoxazole (TMP-SMX) were prescribed. Initially she improved. However, on return home from the trip, she was taken to see a local pediatrician when a nodule developed at the angle of the jaw. Her temperature was 100.9 °F.

She was subsequently admitted to the hospital for 5 days from 4/8 to 4/13/09 for incision and drainage (I&D) of a right submandibular node (a 2 cm area of abscess near the angle of jaw) that did not eventually grow in culture. No histopathologic studies were done on that specimen. A single blood culture drawn on that admission was negative. She received clindamycin and TMP-SMX. During outpatient follow-up on 5/4/09, a new nodule had developed on the right neck, proximal to the old site. Amoxicillin/clavulanate was prescribed, and a TB skin test (PPD) was applied; it was later read as negative. Chest x-ray showed some perihilar markings and with the presence of cough, the possibility of bronchiolitis or pneumonia was raised. An additional antibiotic, azithromycin, was prescribed.

The mother sought a second opinion because the neck nodes and swelling were persisting. A second PPD skin test was applied, described as suspiciously positive, and the patient was admitted on 5/13/09.

Infectious diseases consult was placed on 5/13/09 for help in addressing the cervical lymphadenopathy and the suspected positive PPD skin test.

Review of systems was significant for a relatively healthy-looking child, with no weight loss. She was comfortable as long as the neck was not examined. Her appetite and general well-being were normal. There was no high spiking fever reported. She had no known drug allergies.

Cases in Clinical Infectious Disease Practice: Obtaining a Good History from the Patient Remains the Cornerstone of an Accurate Clinical Diagnosis: Lessons Learned in Many Years of Clinical Practice,
First Edition. Okechukwu Ekenna.
© 2016 John Wiley & Sons, Inc. Published 2016 by John Wiley & Sons, Inc.

The past medical history was significant for a flat congenital hemangioma at the posterior neck area, frequent ear infections in the past, and a recent cough for which chest x-rays were done on 5/4 and 5/11/09. Other history is as noted above. She was said to be up to date on her immunizations.

Epidemiologic history was significant for the following: she was an only child, living with her young parents. Recently, at the time of the initial neck swelling in March, she had been traveling with the parents across the States (father was a truck driver) for about 2 months. However, they slept and cooked in the truck, ate nothing unusual, and did not sleep in hotels. She had no cat bites or scratch; she had in fact no animals or contact with animals. There was no one in the family with known tuberculosis (TB) or contact to TB, including the grandparents. There was no ingestion of unpasteurized milk or dairy products. There was no report of any dental problem or recent surgical procedure.

On examination, she was alert and well oriented, and not in acute distress. Vital signs were as follows: blood pressure (BP) 110/55, respiratory rate (RR) 28, heart rate (HR) 132, and temperature 99.8 °F; height 2′8″, weight 24.4 pounds. Head and neck exam showed no jaundice or conjunctivitis. The oropharynx was only partially examined due to poor co-operation, but there was no obvious mucosal or dental abnormality noted. The left neck was unremarkable, while the right neck showed an old healed scar/scab at the angle of the jaw (previous surgery site on 4/8/09), and a slightly red, non-adherent nodule above the old surgical site. The heart, lung, abdominal, external genitalia, and lower extremities were normal. The right arm had a peripheral intravenous line in place while the left arm showed several punctate lesions, apparently reflecting the two recent PPD skin tests placed in the previous week, and 2 days prior to admission. Neurologic examination was normal, except for irritable cry when the right neck was examined. The skin showed a flat congenital hemangioma at the nape or back of the neck, but no rash or other changes, except at the right neck, as shown in Fig. 13.1a.

• *What are the differential diagnoses that come to mind with this history and picture?*

Laboratory and radiologic findings

The chest x-rays done before admission on 5/4 and 5/11/09 showed perihilar markings in the first x-ray that were largely resolved on 5/11/09. Computed tomography (CT) scan and ultrasound of the neck done on 5/11/09 confirmed prominent lymph nodes in the right neck region, and a 1.2 × 2.3 cm rim-enhancing, low-density collection in the soft tissue suggestive of abscess. The erythrocyte sedimentation rate on various occasions in April and May was normal, ranging between 0 and 4 mm/hour. On 5/11/09, the complete blood count (CBC) showed a white blood cell (WBC) count of 10.4, hemoglobin and hematocrit (H/H) 12.5 and 37.7, respectively, mean corpuscular volume (MCV) 82.5, and platelet count 368,000. The differential count showed 33.7%

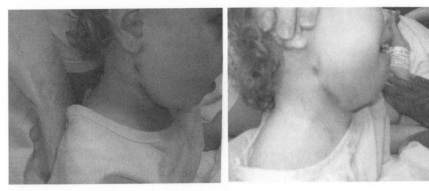

Figure 13.1a Right cervical lymphadenitis. Photo taken on 5/13/09, 1 day before the second surgery (reproduced with permission).

polymorphonuclear leukocytes (PMNs), 53.4% lymphocytes, 8.4% monocytes, and 4.0% eosinophils. Chemistries showed generally normal values: sodium 140, potassium 4.3, glucose 100 mg/dL, blood urea nitrogen (BUN)/creatinine 9/0.4, respectively; serum glutamic oxaloacetic transaminase (SGOT) 41, serum glutamic pyruvic transaminase (SGPT) 15, albumin 3.4, and alkaline phosphatase 211. Lactate dehydrogenase (LDH) was elevated at 331 (normal 98–192 IU/L).

Hospital course and subsequent events

The infectious diseases consultant recommended excisional biopsy of the right cervical lymph nodes, in view of the differential diagnoses considered in this stable patient. This procedure was carried out by the surgeon on 5/14/09. At surgery, a 2.5 × 1.5 cm neck mass (with caseating purulence) was excised. Multiple stains and microbiologic cultures were performed.

• *What do you think the histopathology later showed?*

A lumbar puncture was done showing normal or negative findings. The second PPD skin test was read as likely postive at about 10 mm induration, although the technique of the application was questionable. Blood was sent off for QuantiFERON-TB Gold test, to address the questionable PPD result. The patient remained clinically stable. A graphic display of the vital signs during the hospital admission (5/13–5/18/09) is shown in Fig. 13.1b.

The results of the histopathology on the neck mass were available 24 hours after the biopsy. They showed marked chronic granulomatous inflammation with giant cells and necrosis. Acid-fast bacilli (AFB) were also identified on special stains. No fungal elements were noted on the Grocott's methenamine silver (GMS) stain. The culture results were still pending at this time. The QuantiFERON-TB Gold test and the final culture report of the node would become available only after the patient was discharged home to outpatient follow-up.

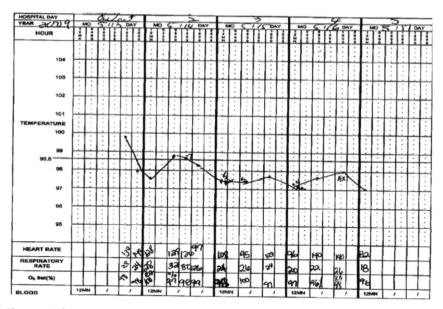

Figure 13.1b Graphic display of vital signs (hospital admission) between 5/13 and 5/17/09.

• *Based on the available data, can you hazard a guess as to the likely diagnosis?*

A presumptive diagnosis was made, and treatment recommendations made to the attending physician.

Follow-up and additional comments

This patient was clinically stable, with no significant systemic complaints. She was sensitive to touch or examination of the right neck, but otherwise was well, had no fever or weight loss, and was eating and playing well for a now 18-month-old child. She had already been through two surgeries, and was understandably sensitive.

The excisional biopsy had yielded useful data, and the clinical diagnosis had been substantiated, even though the final node culture report was still pending. Two options were given to the attending physician: do close follow-up in the outpatient setting on no specific antimicrobial therapy, or begin empiric medical therapy while awaiting the final culture result of the node biopsy. The attending physician chose to begin empiric antimicrobial therapy.

Over the next year, the patient was seen several times in the outpatient setting by the attending physician for other issues but did not require any more in-hospital admissions. Multiple other tests, including CT of the head, serologies to Epstein–Barr virus (EBV), cat scratch disease (CSD), antistreptolysin O (ASO), Monospot, HIV, and C-reactive protein, were all negative. She is said to be doing well, and is fully recovered. The family was screened by the local health department authorities. No TB was found.

Additional laboratory results

The QuantiFERON-TB Gold test was reported positive the week after discharge (a send-out test). The tissue culture was reported positive for AFB within 19 days. The specimen was then sent off to a reference laboratory for Gen probe, to rule out *M. tuberculosis*, and to perform antimycobacterial sensitivity testing.

- *What is your diagnosis now?*

Case discussion

This patient presented with the classic features of chronic cervical lymphadenitis in children. The symptoms lingered for over 2 months before a final diagnosis was made. She was seen initially in March (out of state) for what was clearly not a true "staph" infection, and then required two further admissions for resolution of her problems. The first biopsy was an incision and drainage procedure, which was not curative because she relapsed. During the second admission, we suggested an excisional biopsy, which we think was curative.

The organism finally identified and confirmed by Gen probe was *Mycobacterium avium-intracellulare* complex (MAI), a non-tuberculous mycobacterium (NTM). This was not a surprise diagnosis in this patient, given her relative well-being and the clinical presentation and history.

Infectious causes of cervical lymphadenitis are divided into four general groups: acute unilateral, chronic unilateral, acute bilateral and chronic bilateral [1,2]. The patient presented here would be classified as being in the chronic unilateral group. The most common causes under this group include NTM (also called mycobacteria other than tuberculosis or MOTT), tuberculosis (*M. tuberculosis*), and cat scratch disease [1,2]. Of course, there are many non-infectious causes of cervical adenopathy, but these will not be discussed here.

The epidemiology, history, and physical examination in this patient clearly favored NTM as the cause of the chronic adenopathy. For that reason, the infectious disease recommendation was for excisional biopsy, which is usually curative (96% in one report) [3], as opposed to medical treatment alone (66%) [3]. We did not recommend incision and drainage because of suppuration and sinus tracts that can develop and persist in non-excised lymph nodes, as a complication.

As it turned out, the antimicrobial therapy started presumptively by the attending physician was likely ineffective (since the NTM [MAI] was resistant to those agents: isoniazid, rifampin plus pyrazinamide, which was not on the panel). Clarithromycin was found to be a sensitive agent against this NTM. A typical presumptive therapy for NTM cervical adenitis (targeted at MAI) usually includes clarithromycin and ethambutol; both were sensitive agents in this patient.

The usefulness and interpretation of the new QuantiFERON-TB Gold test in children less than 5 years have not yet been defined [4]. However, the smaller induration of the PPD skin test would be compatible in children with an infection with NTM, typically 5–15 mm, as in this setting [2]. The patient had an induration of about 10 mm with the second PPD skin test.

- *Final diagnosis: Subacute/chronic cervical lymphadenitis secondary to non-tuberculous mycobacterial (NTM) infection (specifically, MAI complex).*

Lessons learned from this case

- The importance of differentiating between the various categories of infectious cervical adenitis, especially in children, as this has clinical and therapeutic implications (acute, bilateral, chronic, and unilateral).
- The preference is for excisional biopsy rather than incision and drainage in addressing chronic cervical adenitis in well children (in this patient, the initial I&D did not lead to a cure, but instead to a recurrence).
- The relative well-being of the patient was a clue that the likely cause was a NTM, as opposed to *M. tuberculosis.*
- The lack of response to multiple antibiotic courses was another clue that this was not a simple acute pyogenic infection, and an indication for excisional biopsy.
- As in this case, it is important to do histopathologic analysis on any excised lymph node (this was not done during the first admission).
- In a young child especially, NTM (MAI or scrofula) is an important differential in subacute/chronic, unilateral cervical adenitis (in adults, *M. tuberculosis* may be more common).

References

1 Healy CM. Cervical lymphadenitis in children: etiology and clinical manifestations. Available at: www.uptodate.com/online/content/topic.do?topicKey=pedi_id/17227&selectedTitle=5 %7E44&source=search_result, accessed March 7, 2016.

2 Healy CM. Cervical lymphadenitis in children: diagnostic approach to and initial management. Available at: www.uptodate.com/contents/diagnostic-approach-to-and-initial-treatment-of-cervical-lymphadenitis-in-children?view=print, accessed March 7, 2016.

3 Lindeboom JA, Kuijper EJ, Bruijnesteijn van Coppenraet ES, Lindeboom R, Prins JM. Surgical excision versus antibiotic treatment for nontuberculous mycobacterial cervicofacial lymphadenitis in children: a multicenter, randomized, controlled trial. Clinical Infectious Disease 2007; 44(8): 1057–64.

4 Guidelines for Using the QuantiFERON®-TB Gold test for detecting Mycobacterium tuberculosis infection, United States. Morbidity and Mortality Weekly Report 2005; 54(RR15);49–55. Available at: www.cdc.gov/mmwr/preview/mmwrhtml/rr5415a4.htm, accessed March 4, 2016.

Further reading

Griffith DE. Pathogenesis of nontuberculous mycobacterial infections. Available at: www
.uptodate.com/online/content/topic.do?topicKey=othr_myc/11705&selectedTitle=9%7
E150&source=search_result, accessed March 4, 2016.

Griffith DE. Overview of nontuberculous mycobacterial infections in HIV-negative patients.
Available at: www.uptodate.com/online/content/topic.do?topicKey=othr_myc/10997,
accessed March 4, 2016.

Griffith DE. Microbiology of nontuberculous mycobacteria. Available at: www.uptodate
.com/online/content/topic.do?topicKey=othr_myc/11028&selectedTitle=1%7E150&
source=search_result, accessed March 4, 2016.

QuantiFERON®-TB Gold Fact Sheet. New York City Department of Health and Mental Hygiene,
Bureau of Tuberculosis Control. Available at: http://www.nyc.gov/html/doh/downloads/pdf/
tb/tb-qtf-factsheet.pdf, accessed March 4, 2016.

Case 13.2   Persistent headache in a 50 year old with known migraine

Three days before admission, a 50-year-old Caucasian female presented to the emergency room (ER) with acute onset of diffuse headache, similar to previous migraine. She received treatment with intramuscular injection of meperidine and promethazine, and a prescription for ondansetron for nausea. She returned the next day with worsening headache (described as 10/10 in intensity), still nauseated, but she had no fever or chills (temperature 98.2°F). CT scan of the head was unremarkable. She was asked to return if there was no improvement in symptoms.

Two days later on 7/6/09, the patient returned to the ER with persisting and severe headache that had not been relieved by sumatriptan and previous medications, including narcotics. Fever had developed the night before, along with stiff neck, and friends had advised about possible meningitis.

In the ER, a lumbar puncture was done, and the patient was admitted.

On 7/7/09, 1 day after admission, infectious diseases consult was placed for "meningitis" (viral or bacterial?).

The review of systems was significant for the following: diarrhea (≥3 BM/day), lasting 5–6 days and resolving 1–2 days prior to admission (PTA); severe headache and neck stiffness ×3 days PTA, unresponsive to treatment for migraine; fever starting 1 day PTA; nausea and poor appetite; migraine headache, an old problem for years; photophobia (she preferred a quiet and calm environment); and no significant body aches.

The past medical history was significant for chronic migraine headaches for years; hypertension; hypercholesterolemia; carpal tunnel surgeries 3 years earlier; degenerative joint disease (osteoarthritis) of the back, hands, and chronic low back pain; plus a history of some anxiety. Medications prior to admission included olmesartan; valsartan; combination headache pill –butalbital/acetaminophen/caffeine; sumatriptan; atropine/diphenoxylate for diarrhea; pravastatin; metoprolol; hydrocodone/acetaminophen; and alprazolam.

The family and social history was pertinent for the following: she was a married "homemaker"; lived in a neighboring state; and was the mother of three adult children (age range 21–30 years). She drank occasional wine, but did not smoke. She was allergic to "sulfa" (mouth blisters).

Epidemiologic history was relevant for the following: her husband was healthy, and working out of state (where she had recently joined him, visiting in the past week). She enjoyed gardening a lot and being outdoors, and had a good suntan from exposure to sunlight. She also enjoyed eating soft cheese (including Mexican cheese) and other dairy products. They had three healthy indoor dogs. There was no skin rash or petechiae or any obvious cuts on the body or skin.

The physical examination showed her to be oriented but uncomfortable; lying calm and quiet in bed, and still complaining of headache and nausea. The vital signs on admission were as follows: BP 168/87, RR 16, HR 93, temperature 101.3 °F; height was 5'5" and weight 158.6 pounds. The oxygen saturation was 96% on room air. The next day, on 7/7/09, the BP was 99/55, RR 16, HR 64, and temperature still elevated at 101.5 °F. Earlier, at 0300 hours, the temperature had reached a peak of 104.6 °F, probably precipitating the ID consultation.

The head and neck exam showed no evidence for conjunctivitis or jaundice. The teeth were in good condition and the oral mucosa normal. The heart, lung, and abdominal exam showed no abnormalities. Lymph nodes were not palpably enlarged in the neck, axilla, or groin. The upper and lower extremities were normal, except for degenerative changes of the joints of hands and knees. The skin showed a good suntan in sun-exposed areas. Neurologic exam showed a patient who was somber, had slightly depressed affect, with no focal findings, but diminished deep tendon reflexes at the knees. Cranial nerves were intact while stiff neck previously noted on 7/6/09 was unresolved, but not worse.

Laboratory and radiologic findings

Non-contrast cranial CT scan 2 days before admission was read as negative. On 7/6/09, the day of admission, the chemistries showed a sodium value of 126, potassium 2.6, glucose 152 mg/dL, and BUN/creatinine 12 and 0.6, respectively. SGOT was 25, SGPT 47, alkaline phosphatase 59, total bilirubin 0.5, albumin 3.2, and total protein 7.1. CBC showed a WBC count of 15.0, H/H 12.9/37.5, respectively, MCV 91.5, and platelet count 306,000. The WBC differential counts were as follows: 81% PMNs, 9% lymphocytes, and 9.6% monocytes.

Cerebrospinal fluid (CSF) findings included a total WBC of 2138/mm³ and red blood cells (RBCs) 10; differential count: 40% PMNs, 55% lymphocytes, and

5% mesothelial cells. Glucose was 43 mg/dL, and protein 107 mg/dL. The gram stain showed a few WBCs and no organisms. The CSF bacterial antigen panel was negative. One day after admission, the blood cultures were reported negative.

Early events in the hospital course

In the ER on 7/6/09, vancomycin and ceftriaxone were initiated after lumbar puncture was done, on the presumptive diagnosis of meningitis. The patient continued to have fever with chills, headache, and stiff neck. The temperature reached a height of 104.6 °F at 0300 hours on 7/7/09. Nausea and vomiting persisted, several times per day, even as I was examining her on 7/7/09. When the microbiology lab was called, they confirmed that turbid growth was noted in the CSF broth culture, and also a tiny creamy growth on the primary culture plate, perhaps with slight hemolysis noted around the tiny colonies. The CSF gram stain (broth and primary plate) showed gram-positive coccobacilli or diphtheroids.

- *Based on this preliminary information, what are the likely differential diagnoses?*
- *What treatment modifications would you make and why?*

A decision was made at this time to make some modifications in treatment.

On 7/7/09, the day of the consult, and based on the preliminary information from the gram stain, modification in antimicrobial therapy was made (details later). Fever continued, up to a temperature of 103.9 °F at 0030 hours on 7/8/09. Later that same day, the CSF culture was confirmed positive for a gram-positive organism (48 hours after lumbar puncture and CSF culture).

The blood agar (BA) and trypticase soy broth (TSB) cultures of the CSF are shown in Fig. 13.2a. The organism was strongly beta-hemolytic on BA (not clearly shown here, for patient privacy reasons).

- *Can you name the likely agent and why?*

On day 3 after admission, the patient was clinically slightly better, temperature was now normal and vital signs more stable: BP 120/69, RR 18, HR 71, temperature 96.5 °F. She was able to take a shower, the headache was better, and she had no diarrhea.

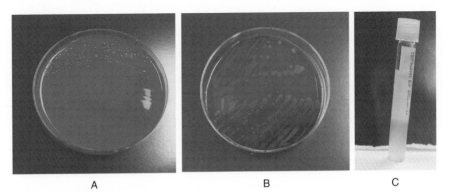

A   B   C

Figure 13.2a Blood agar plate cultures: A. Day 1. B. Day 9. C. Broth culture of CSF on day 9. (*See insert for color information*)

On day 3, the single blood culture drawn on 7/6/09 was reported positive. The laboratory was not sure of the significance of this since it was a "diphtheroid-like" organism. Additional blood cultures were ordered which turned out subsequently to be negative.

- *Assuming that the initial blood and CSF cultures were accurate, what would be your microbiologic diagnosis of the meningitis in this patient?*

Case discussion

Following the initial consult of 7/7/09, the CSF findings and the epidemiology suggested a need to modify therapy. A combination of ampicillin (2 g Q 4 hours) and gentamicin (80 mg Q 8 hours) was added to the vancomycin therapy, and ceftriaxone was discontinued.

- *What was the thinking of the ID consultant in making these treatment changes?*

The CSF culture was confirmed the next day. Magnetic resonance imaging (MRI) of the brain on 7/10/09 showed no abscess.

- *Why would we be concerned about brain abscess in this patient?*
- *List the possible differential organisms for gram-positive coccobacilli in CSF in a patient with meningitis.*

The likely organisms would include the possibility of a contaminant (diphtheroid) or a true infection. Common and typical causes of meningitis in adults in this setting would include pneumococcus (*Strep. pneumoniae*), gram-variable *Haemophilus* spp, or *Listeria monocytogenes*.

However, with small zones of beta-hemolysis on the blood agar plate, the differential shifts toward *Listeria monocytogenes* as the most likely agent for this meningitis.

This organism was subsequently found to be sensitive to penicillin/ampicillin.

- *Final diagnosis: Listeria monocytogenes meningitis and bacteremia.*

Listeria monocytogenes

Listeria is an aerobic and facultatively anaerobic, motile, beta-hemolytic, non-spore-forming, short, gram-positive rod that exhibits characteristic tumbling motility on light microscopy [1]. *Listeria* occurs singly or in short chains. On gram stain, it may resemble pneumococci (diplococci), enterococci or diphtheroids (Corynebacteria), or be gram variable and be confused with *Haemophilus* species. The organism produces a characteristic appearance on blood agar with small zones of clear beta-hemolysis around each colony (Fig. 13.2b).

Gelfand has outlined some characteristics of the organism, including habitat and pathogenesis [1]. *Listeria* is a facultative intracellular parasite that grows well even at refrigeration temperatures (4–10 °C). The primary habitat of *Listeria* is the soil and decaying vegetable matter. Most infections in adults are thought to result from oral ingestion and subsequent intestinal mucosal penetration and systemic infection.

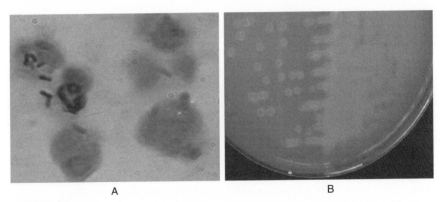

A   B

Figure 13.2b A. Gram stain of cerebrospinal fluid showing inflammatory cells and small, gram-positive rods and coccobacilli. B. Clear areas around each colony of *L. monocytogenes* are characteristic small zones of beta-hemolysis on blood agar [1]. Reproduced with permission from: Waltzman M. Initial evaluation of shock in children. Available at: www.uptodate.com/contents/initial-evaluation-of-shock-in-children, accessed on March 4, 2016. Copyright © 2015 UpToDate, Inc. For more information visit www.uptodate.com. (*See insert for color information*)

In the United States, the incidence of laboratory-confirmed cases of listeriosis in 2009 was 0.34/100,000 persons [2,3]. It has the second or third highest mortality rate (20%) among food-borne infections in the US. The organism has also been isolated from dust, numerous human food products, animal feed, water, sewage, numerous species of animals, and asymptomatic humans. *Listeria* has a predilection for the placenta and central nervous system, crossing the placenta during maternal bacteremia and infecting the fetus [1,4].

Gelfand has reviewed the clinical syndromes of listeriosis, and noted that invasive disease may manifest in several forms: infection in pregnancy; sepsis of unknown origin; central nervous system infection (meningitis, meningoencephalitis, cerebritis, rhombencephalitis); focal infections; and neonatal infections [5]. In normal hosts who ingest high numbers of organisms, it can cause a self-limited febrile gastroenteritis.

Epidemiology

Most systemic, invasive *Listeria* infections occur in individuals with one or more predisposing conditions. These include pregnancy, glucocorticoid therapy, other immunocompromising conditions, and age. Glucocorticoid therapy is the most important predisposing factor in non-pregnant patients. *Listeria* is a common low-level contaminant of both processed and unprocessed foods of plant and animal origin. The most commonly reported foods in outbreaks are processed or delicatessen meats, hot dogs, soft cheeses, smoked seafood, meat spreads, and patés [2,3]. Most of the reported infections, however, are sporadic (95%).

- *What is the recommended duration of therapy for listeriosis?*

Gelfand suggests the following approach to managing patients [5]. In immuno-competent patients, 2 weeks is sufficient for bacteremia and 2–4 weeks for central nervous system (CNS) infection. Relapses have occurred in immuno-compromised patients after 2 weeks of treatment. As a result, 3–6 weeks is preferable in patients with bacteremia and 4–8 weeks in those with CNS infections. The longer duration particularly applies to patients with cerebritis or brain abscess. When given, gentamicin is continued until the patient improves (usually 7–14 days) or, in poor responders, for up to 3 weeks if there are no signs of nephrotoxicity or ototoxicity. Ampicillin or penicillin is the usual drug of choice. Trimethoprim-sulfamethoxazole is a good alternative in patients allergic to penicillin (especially where penicillin desenitization is not possible).

Postscript

In this patient, the illness started with diarrhea 1 week prior to admission. Diarrhea lasted for 5–6 days, and had stopped at the time of admission. It is likely that the pathogen was orally ingested in food (which specific food is unclear).

Several food sources have been implicated in outbreaks (see Epidemiology section). The patient was not overtly immunosuppressed, and was not on glu-cocorticoid therapy. By the time the meningitis was diagnosed, the diarrhea had resolved.

On 7/13/09, after 7 days in hospital, the patient was transferred to her home state, where she completed the planned 3-week therapy with assistance from Home Health. She was not considered immunocompromised.

When contacted 2 years later by phone, she confirmed that she was still doing well, and appeared to have had no major sequelae from her illness.

Although we have no proof for the source of her infection in 2009, she now avoids soft cheese, as one possible source of that infection.

Lessons learned from this case

- Knowledge and competence in reading and interpreting the gram stain can be crucial in making a timely diagnosis or influencing therapy (the treatment was adjusted on the day of the ID consult primarily based on the findings of the gram stain).
- The single blood culture (BC) obtained on admission (instead of two sets) nearly led to the false assumption that the blood culture report may have reflected a contaminant organism ("diphtheroids" are notorious laboratory contaminants). The standard practice is at least two sets of BCs.
- This single blood culture was not dismissed as a contaminant, especially because it was hemolytic on blood agar, just like the CSF culture, and had other characteristics of *Listeria* (such as positive motility).
- Brain MRI was done to rule out brain abscess because of persistent headaches, and because that is a known complication of *Listeria* meningitis. It was also

important to know since that finding would have affected the duration of antimicrobial therapy (extended therapy with brain abscess as noted above).

- Follow-up of patients after completion of therapy is useful to confirm that there are no complications. (This patient remained well more than 2 years after her illness.)

References

1 Gelfand MS. Epidemiology and pathogenesis of Listeria monocytogenes infection. Available at: www.uptodate.com/contents/epidemiology-and-pathogenesis-of-listeria-monocytogenes-infection?view=print, accessed March 4, 2016.

2 CDC. Outbreak of invasive listeriosis associated with the consumption of hog head cheese – Louisiana, 2010. Morbidity and Mortality Weekly Report 2011; 60(13): 401–5. Available at: www.cdc.gov/mmwr/preview/mmwrhtml/mm6013a2.htm?s_cid=mm6013a2_w, accessed March 4, 2016.

3 CDC. Preliminary FoodNet data on the incidence of infection with pathogens transmitted commonly through food – 10 states, 2009. Morbidity and Mortality Weekly Report 2010; 59(14): 418–22. Available at: www.cdc.gov/mmwr/preview/mmwrhtml/mm5914a2.htm, accessed March 4,2016.

4 Gelfand MS.Clinical manifestations and diagnosis of Listeria monocytogenes infection. Available at: www.uptodate.com/contents/clinical-manifestations-and-diagnosis-of-listeria-monocytogenes-infection?view=print, accessed March 4, 2016.

5 Gelfand MS. Treatment, prognosis, and prevention of Listeris monocytogenes infection. Available at: www.uptodate.com/contents/treatment-prognosis-and-prevention-of-listeria-monocytogenes-infection?view=print, accessed March 4, 2016.

Case 13.3   A 54-year-old man with skin lesions with central numbness

A 54-year-old Caucasian male was referred for chronic skin lesions following a biopsy.

In late September 2010, he was seen in the office, accompanied by his wife. New skin lesions had appeared, beginning at least 6 months earlier, initially in the proximal left arm and elbow area, with several other lesions in other body sites in the subsequent 3 months. The skin lesions got larger over time.

The review of systems showed that the patient did not feel ill; he looked healthy. He had no itch or pain, but anesthesia at the center of the flat older skin lesions. There were no chills or fever, no body aches, except for old arthritis of the shoulders. He had no skin abscesses, no nasal discharge, and no cough.

The past medical history was significant for recurrent impetigo in the upper and lower extremities over the past 2–3 years (these are different from the newer skin lesions). He had been under the care of a dermatologist for 2 years, with mainly topical agents prescribed for the impetigo. Two skin punch biopsies done by the dermatologist on the left forearm on 3/30/09 showed the following from the two sites: suppurative folliculitis and impetiginized non-specific ulcer. No fungal elements were found on the biopsies. The patient was treated over several months with topical and systemic antibiotic regimens, antifungal, and topical steroid therapy, without much effect. He had arthritis of the shoulder and elbows for years. No other serious medical or surgical problem was reported by the patient.

A biopsy of one of the new flat skin lesions was done on 9/10/10, following which he was referred to the office on 9/29/10.

The epidemiologic history suggested that he did a lot of vocational work in the yard, did woodwork, worked with and had contact with lawn, soil, and grass. He had bought an old, previously abandoned house, with overgrown yard and lawn, several years earlier. He rebuilt this house over a 10-month period, including cleaning, clearing, plumbing work, and crawling under the elevation. He had two healthy dogs (an outside dog, aged 7 years, and an inside dog 3 years old). He had been married to his wife for 22 years. She was healthy and had no skin lesions.

The family and social history included a daughter, aged 20 years, who was healthy, and lived at home with the parents. The patient worked as a firefighter. He did not smoke and very rarely drank alcohol. He had no drug allergies. He was on no regular medications, except for PRN tramadol and naproxen for arthritis, but was recently prescribed alpraxolam (0.25 mg) PRN, for sleep and anxiety.

On examination, he was alert and oriented, slightly anxious, but in no acute distress. The vital signs were as follows: BP 130/80, RR 20, HR 76, temperature 98.1 °F; height 5'9'' and weight 161 pounds. The head and neck exam was unremarkable. He had normal oral mucosa, several tooth fillings, and no lesions in the face, nose, mouth, or neck. The heart, lung, and abdomen were normal. The external genitalia showed a normal circumcised male phallus. The lymph nodes were not enlarged in the neck, axilla, or groin. The neurologic exam was normal, except for the skin lesions. The upper extremities were unremarkable, except the skin lesions to be discussed below. The lower extremities showed mild changes suggestive of tinea pedis between the fourth and fifth toes, but otherwise were normal.

The skin findings were as follows. The left elbow radial forearm showed a large lesion with a flat center that was dry and anesthetic; the extensor area just above the right elbow showed a discoid, eczematous area with distinct borders. There were also a few younger lesions noted in the left and right flanks, as well as inner and outer elbow areas. The right posterolateral flank lesion was about 3–4 cm in size, raised, oval shaped or elliptical, and purplish. Note: the oldest lesion was in the radial left elbow; the biopsied site was one of the younger lesions, in the right upper arm extensor elbow area. A sketch of the left forearm lesion is shown in Fig. 13.3a.

Figure 13.3a Sketch of left proximal extensor forearm skin lesion. The 5–6 cm lesion showed an anesthetic flat center, with surrounding raised borders (adapted from a photograph taken in the office on 9/29/10).

- *What are the differential diagnoses that come to mind, based on the data available so far and the physical examination findings?*

Results of preliminary investigations

The chest x-ray done on 9/29/10 was normal. Chemistries done on 10/1/10 showed the following: sodium/potassium 141/4.5, respectively, glucose 93 mg/dL, BUN/creatinine 16/1.1, respectively. Liver function tests and lipid profile were normal. The CBC showed a WBC of 6.3, H/H 12.6/39.7, respectively, MCV 61.6, and platelet count 140; differential: 71 P, 21 L, 7 M, 1 eosinophil. HIV serology was negative; glucose-6-phosphate dehydrogenase (G6PD) level 18 U/g hemoglobin (normal range 7–20).

Case discussion

First, let us discuss the character of the skin lesions. These were typically large (several centimeters) and flat, with numbness developing in the center as the lesions got older. There was no itch or pain associated with the lesions. The lesions became flatter and dry in the center with aging, and were associated with loss of hair. There were no lesions on the face or mouth area. The lesions were noted in bilateral upper extremities, especially the elbow, and also flank areas.

- *Do these additional descriptions influence your differential diagnoses?*

Our initial differential diagnoses before review of the biopsy included the following.

- Sarcoidosis
- Annular psoriasis
- Non-specific dermatitis
- Atypical mycobacterial infection
- Deep fungal infection
- Subcutaneous mycosis (e.g. dematiaceous fungal infection)
- Systemic lupus erythematosus
- Cutaneous leishmaniasis
- Leprosy (*Mycobacterium leprae*)

The histopathology of the skin punch biopsy of the right upper arm done on 9/10/10, a biopsy of skin to the level of the subcutis, was described as showing non-necrotizing granulomatous inflammation involving the dermis with numerous acid-fast-positive organisms.

- *How do the new findings modify your differential diagnoses?*

More detailed characterization of the histopathologic findings revealed the following in the histopathology report:

> The epidermis exhibited hyperkeratosis and focal acanthosis. Portions of the superficial dermis display fibrosis consistent with scar. In the dermis there are focal infiltrates of non-necrotizing granulomatous inflammation admixed with a moderate lymphocytic inflammatory infiltrate. In focal areas, inflammatory infiltrate displays a perineural distribution. Numerous AFB are present in Fite-stained sections. No AFB or fungal organisms are identified in either routine AFB or PAS-stained sections, respectively. No foreign material is identified under polarized light.

- *What is your final diagnosis?*

The final histopathologic diagnosis was noted by the dermatopathologist as follows:

> The histopathologic findings show granulomatous inflammation and numerous acid-fast-positive organisms in Fite-stained sections, consistent with a mycobacterial infection. Given the focal perineural pattern of inflammatory infiltrate, positive staining of numerous organisms in Fite-stained sections, and absence of staining in routine AFB stained sections, the findings are suggestive of the lepromatous type of leprosy.

- *Preliminary diagnosis: Leprosy (Mycobacterium leprae infection).*

Follow-up findings and more data on the patient

Following extensive discussions with the patient and review of the epidemiology, as well as review of the histopathology of the skin biopsy done earlier, a tentative diagnosis was made. A call was made to a specialized center to discuss this diagnosis. The patient suggested (and I agreed) that he should be referred to this tertiary center, specializing in Hansen's disease. A decision was then made to refer the patient to this center for further work-up, treatment, and follow-up.

Follow-up at the Hansen's disease center

Work-up for definitive diagnosis and treatment was initiated with the National Hansen's Disease Programs (NHDP) in Baton Rouge, LA. At the NHDP, the patient had additional skin biopsy just above the right elbow and skin smears performed from typical sites, which showed high bacterial index in some places and negative findings in other places.

Treatment with three drugs was started, with follow-up scheduled every 3 months. Laboratory tests to be followed included CBC and liver function tests. His treatment was scheduled to last for 24 months.

Months later, during a follow-up visit, the diagnosis of thalassemia minor was confirmed when the patient became symptomatic with shortness of breath while on treatment (he was then found to be severely anemic). The anemia was thought to be secondary to dapsone, even though the screening G6PD level was normal. The dapsone was stopped and replaced with clarithromycin, with the new three-drug regimen being rifampin, clofazimen, and clarithromycin.

The patient has completed a 24-month treatment program, according to the NHDP protocol. Follow-up by this center will still continue for years.

Epidemiology of leprosy and *Mycobacterium leprae* infection in this patient

This patient was relatively healthy and looked well. He had multiple skin patches (≥ 6 lesions), with central areas of anesthesia. His symptoms had been ongoing for at least 6–12 months or longer, although there was some concern initially that he may have had bacterial impetigo in 2009. At that time the punch biopsy showed suppurative folliculitis and impetiginized non-specific ulcer.

One of the key epidemiologic risk factors identified in this patient was that he was exposed to an armadillo-infested environment over a prolonged period, while renovating his home. He confirmed that the grounds of the abandoned old building that he bought were infested with armadillos. The renovation included crawling under buildings, doing plumbing work, and other activities that led to prolonged contact with contaminated soil and armadillo-infested grounds. It is thought that this was the most likely way in which he may have acquired the infection.

The literature suggests that currently, most cases of leprosy worldwide are from S.E. Asia (India, Indonesia, Bangladesh), followed by the Americas (e.g. Brazil), then Africa (e.g. Nigeria) [1,2]. However, these data are based mostly on passive case finding and not active case-finding reports. Between 1985 and 2010, the number of registered cases fell from 5.4 million to 244,796, while the prevalence rate per 10,000 fell from 21.1 to 0.37.

In the United States, endemic cases have been identified in New York City, Louisiana, Texas, and Mississippi. According to the NHDP, 205 new cases were detected in the US in 2010 [1]. About 6500 cases were registered in 2006 with the NHDP, the vast majority of these being in Hispanic and Asian immigrants; 75% of new US cases annually are among immigrants. Some cases in native-born US citizens may be due to overseas exposure, and some to exposure to infected armadillos. In some cases no history of exposure can be established. In North America (USA), leprosy appears sometimes to be a zoonosis (specifically from armadillos).

Our patient most likely acquired his leprosy locally through direct skin or mucous membrane contact to secretions or tissue from armadillos that infested the grounds of his old house.

Leprosy is a chronic disease that produces sores on the skin and mucous membranes, and infects nerves, producing loss of sensation in the affected areas [1,3]. Clinically, the disease appears as two types. In the more virulent form, or lepromatous leprosy, the numerous sores contain many bacteria; in the milder form, or tuberculoid leprosy, fewer sores appear because of the body's immune response to the infection. Signs of the disease often do not appear for many years. While leprosy is rarely fatal, it can cause permanent disfigurements.

Leprosy in armadillos [3]

Nine-banded armadillos (*Dasypus novemcinctus*) are naturally infected with *Mycobacterium leprae* and have been implicated in zoonotic transmission of leprosy [3]. Studies have found infected armadillos in wide-ranging areas across the south-central regions of the US, especially Texas and Louisiana [3]. Since the leprosy bacillus concentrates in the body's extremities (e.g. fingers, ears) in people, scientists have suggested that the armadillo's susceptibility to leprosy was very likely due to its unusually low body temperature of 90 °F (33 °C), which is more than 8° lower than the temperature of humans and other mammals.

Although the risk of transmission to humans appears low, scientists are still not sure how leprosy is spread to the human population from this animal. However, armadillos can shed leprosy bacilli into the environment in bodily secretions, and bacilli might survive extracellularly in the environment for short periods [3]. Prolonged contact with contaminated environments may therefore be relevant to the acquisition of leprosy from the armadillo.

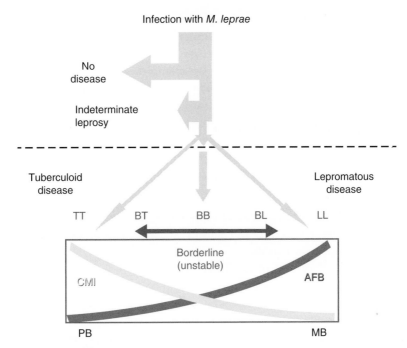

Figure 13.3b The different clinical classifications of leprosy using both the World Health Organization and the Ridley–Joplin systems. The increase in number of acid-fast bacilli and defects in cell-mediated immunity are represented in the continuum from paucibacillary (PB) to multibacillary (MB) disease. Adapted with permission from Elsevier from Jacobson and Krahenbuhl [4].

Classification of leprosy

Leprosy can be classified into the following categories [1,4]: tuberculoid (TT), borderline tuberculoid (BT), mid-borderline (BB), borderline lepromatous (BL), lepromatous (LL), and indeterminate (I) (Fig. 13.3b).

The classification system noted here in the diagram is a practical adaptation of the World Health Organization and the Ridley–Joplin system, combined. There are also other methods of classification of leprosy [5].

Patients with lepromatous leprosy (LL) are thought not to be immuno-compromised, but to have a selective inability to mount cellular immunity to *Mycobacterium leprae* [1].

Final comments

Based on the number of skin lesions, as well as the presence of multiple AFBs seen on histopathology, the working presumptive diagnosis is multibacillary (MB), as opposed to paucibacillary (PB) leprosy.

It is my belief that the patient acquired the *Mycobacterium leprae* infection through contact with armadillo tissue or secretions in the soil and environment of the abandoned old house that was previously infested with armadillos.

Even though he has completed the 24-month treatment program with the NHDP, he is still required to be followed by the NHDP for several years.

Lessons learned from this case

- Leprosy is still endemic in the United States, although very rare.
- Armadillos in south-central United States can be a source of zoonotic leprosy under certain conditions.
- Indigenous leprosy is possible in the native-born US citizen who did not acquire the disease from overseas travel; leprosy can be acquired locally.
- Multiple skin lesions with central anesthesia in the appropriate setting should trigger a suspicion for leprosy.
- A thorough history and physical examination are crucial in initiating the work-up, and arriving at an accurate diagnosis of this chronic but treatable condition.
- The patient presenting with leprosy skin lesions typically does not look ill.
- Preferably, such a patient should be referred to a specialized center experienced in the management of leprosy, as was done in this case.

References

1 Scollard D, Stryjewska B. Epidemiology, microbiology, clinical manifestations, and diagnosis of leprosy. Available at: http://www.uptodate.com/contents/epidemiology-microbiology-clinical-manifestations-and-diagnosis-of-leprosy?view=print, accessed March 4, 2016.

2 Renault CA, Ernst JD. *Mycobacterium* leprae. In: Mandell GL, Bennett JE, Dolin R (eds) *Principles and Practice of Infectious Diseases*, 7th edn. Philadelphia: Churchill Livingstone, 2010, pp. 3165–76.

3 Sharma R, Singh P, Loughry WJ, et al. Zoonotic leprosy in the Southeastern United States. Emerging Infectious Diseases 2015; 21(12): 2127–34. Available at: www.ncbi.nlm.nih.gov/pmc/articles/PMC4672434/, accessed March 6, 2016.

4 Jacobson R, Krahenbuhl JL. Leprosy. Lancet 1999; 353(9153): 655–60.

5 Ridley DS. Histological classification and the immunological spectrum of leprosy. Bulletin of the World Health Organization 1974; 51(5): 451–65. Available at: www.ncbi.nlm.nih.gov/pmc/articles/PMC2366326/pdf/bullwho00471-0017.pdf, accessed March 4, 2016.

Further reading

Scollard D, Stryjewska B. Treatment and prevention of leprosy. Available at: www.uptodate.com/contents/treatment-and-prevention-of-leprosy?view=print, accessed March 4, 2016.

Case 13.4   Delayed diagnosis in a young woman with migratory joint pain

A 23-year-old black female was admitted in early September, 2015 through the emergency department (ED) because of left hip pain and inability to get around.

She was in her usual state of health until 2 weeks prior to this admission. At that time, she developed an acute sharp pain in the medial aspect of the left big toe. She went to a local ED for evaluation, and she was thought to have gout. No aspiration of the joint was done but she was kept overnight. The big toe was red and swollen, but there was not much effusion noted.

Because she was pregnant (around 21 weeks at the time) there was a reluctance to prescribe pain medications other than acetaminophen for relief. She did, however, receive a short course of oral steroid therapy.

Two days later, the pain moved to the left knee although she did not have much swelling. She returned to the same ED and again was told it was probably gouty arthritis. No aspiration was done. A few days later, or 4 days after onset of her illness, her left hip became a problem, with significant pain, such that she had difficulty getting around. She went to another ED on 8/29 and again on 9/1/15 for the left leg pain and left hip pain, respectively. She received symptomatic therapy with steroid or hydrocodone-acetaminophen, and was advised to follow up with her obstetrician or orthopedic specialist.

Over the next several days, she was unable to walk, with the left hip pain becoming unbearable, very intense and sharp (10/10 in intensity), so she returned to the ED on 9/6/15. She was found to have elevated inflammatory markers of erythrocyte sedimentation rate (ESR) and C-reactive protein (CRP). This time she was admitted for further work-up.

She had been prescribed clindamycin by mouth for several days 2 weeks earlier during the initial ED visit elsewhere. But she had been off this medication for about 7 days by the time she was admitted on 9/6/15.

Aspiration of the left hip joint was done 1 day after admission, but the results were pending, including the crystal examination. Septic arthritis was suspected based on initial joint fluid parameters, but there was no definite diagnosis, and the cultures were negative on day 1.

Infectious diseases consult was placed 2 days after admission to address the cause of the septic arthritis.

The review of systems was significant for no fever associated with this illness. She had localized swelling and significant pain in the left big toe MTP joint, but apparently no visible effusion; she developed subsequent pain in the left knee, later migrating to the left hip 4 days after onset of the initial big toe symptoms. She had difficulty walking because of intense (10/10) pain in the left hip.

At the time of the ID consult, 2 days after admission, she still had no fever; blood cultures ×2 were negative on day 2 and left hip aspirate cultures were negative on day 1. The result of the crystal examination for gout was still pending.

She had been placed on intravenous vancomycin to cover for the presumed joint infection.

- *What are the possible differential diagnoses that come to mind at this time?*

The past medical history was significant only for the following: a history of migraine headaches for 10 years, although no attacks in 9 months; she was 21 weeks pregnant and has since had a normal vaginal delivery of a now 2-year-old and healthy son. There were no previous major surgical procedures.

There was nothing significant in her family history. However, there was a questionable history of gout in the father and a family history of hyperlipidemia in her parents. Until a few weeks earlier, she had worked at a local casino. She neither smoked nor drank. She had no known drug allergies.

On examination, she was alert and oriented, and in no acute distress. The vital signs were as follows: BP 122/55, pulse 103, oral temperature 98.4 °F, RR 20; height 5′7″, weight 153 pounds. O_2 saturation on room air was 99%.

The head and neck, heart, and lung examinations were unremarkable. There was a protuberant abdomen from 21 weeks of pregnancy, with nothing else significant. Lymph nodes were not palpably enlarged in the neck, axilla, or groins. There was some tenderness in the left groin but not from enlarged lymph nodes.

The upper extremities were unremarkable. She had artificial greenish painted fingernails. The hands and fingers were otherwise unremarkable, with no signs of joint swelling. The lower extremities showed the left big toe that was no longer swollen but had a flat brownish fleck or spot, 5 mm in size, reflecting a previously "bruised" skin area. The shins showed no edema. The knee showed no obvious swelling. The left hip area was tender, especially in the medial aspect all the way to the anterior lateral region. The joint movement was limited because of discomfort. I was unable to do any further hip maneuvers in this pregnant lady.

No pelvic or rectal examination was done.

The skin showed the following relevant findings: she had several tattoos on the upper back, chest, and extremities, including especially the right thigh. The skin was otherwise healthy with no rash noted. The left medial big toe area had a flat brownish skin fleck as mentioned, about 5 mm in size, over the medial MTP joint area, and without joint swelling at the time.

The neurologic exam showed the following: cranial nerves were intact. She was limited in the movement of the left leg because of pain and discomfort. She had bilateral very brisk deep tendon reflexes at the knees, grade 3–4/5, without sustained clonus at the ankles. Gait was not tested because of patient discomfort. Affect was friendly and appropriate. There was no confusion or hallucination.

Comparative photographs of the left foot and big toe over time are shown in Fig. 13.4a.

- *What differential diagnoses would you consider with the available information so far?*
- *What would you recommend further in the work-up of this patient?*

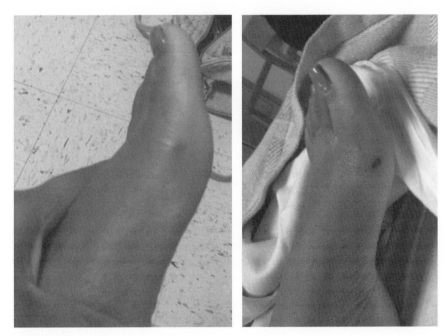

Figure 13.4a Left foot and big toe showing evolving inflammation of the MTP joint. The photo on the left was taken on 8/24/15 by the patient; the photo on the right was taken 16 days later, on 9/9/15, at the hospital (reproduced with permission).

Initial laboratory, radiologic, and other investigations

Previously on 9/1/15, two left hip x-rays, one in anteroposterior (AP) and the other in frog-leg position, showed no evidence of left hip bone or joint abnormalities. MRI of the left hip without contrast, done on 9/7/15, showed a prominent left hip joint effusion with associated adjacent soft tissue edema.

Ultrasound-guided aspiration of the left hip was done by the radiologist on 9/7/15. Approximately 18 mL of cloudy greenish-yellow fluid was aspirated and sent to the lab for analysis (Table 13.4a). MRI lumbar spine without contrast, done on 9/6/15, was unremarkable. The MRI of the cervical spine showed no central canal or foraminal stenosis. The cord showed normal signal.

Blood cultures ×2 on 9/6/15 were negative on day 2. The left hip joint aspirate on 9/7/15 was negative on day 1. ESR was markedly elevated (120 mm/hour) on 9/7/15 (N <20), and CRP was also very high (12.90 mg/dL) on 9/6/15 (N <0.30).

The CBC on admission showed a WBC count of 15.1 × 10*3/uL, with H/H 10.4 and 33.0, respectively; MCV was slightly low at 79.1 fL; and platelet count 452 × 10*3/uL. The chemistries showed nothing significant, except a low albumin of 2.7 g/dL (N 3.4–5.0). The liver panel was otherwise normal, as were the electrolytes: sodium and potassium levels, as well as calcium and magnesium levels.

Table 13.4a Cell count and differential of the left hip aspirate on 9/7/15.

Left hip aspirate	Reference range	9/7/15
WBC fluid	No range found	59,932
RBC fluid	No range found	1081
Neutrophil %	No range found	80
Lymphocyte %	No range found	14
Mono/macrophage %	No range found	6

RBC, red blood cell; WBC, white blood cell.

Hospital course

The differential diagnosis of inflammatory arthropathy, the working diagnosis, was considered wide. The patient was going to be seen by the rheumatologist on that same day, and we were expecting a reading of the left hip aspirate for birefringent crystals, as well as awaiting the final cultures of the left hip aspirate of 9/7/15. The blood cultures were known to be negative on day 2.

The patient was clinically stable and afebrile. No new changes or recommendations would be made for another 24 hours.

• *Would the differential diagnosis change or be narrowed in a pregnant young woman with migratory arthritis, as in this case?*

Additional information was obtained the next day on 9/9/15. On further investigation, it was noted that the patient had not been sexually active since she became pregnant, at least for more than 4 months (she was 21 weeks pregnant at the time of admission). She had not had a formal speculum pelvic exam by her obstetrician and gynecologist (not usually done at this stage of pregnancy); and she had no vaginal discharge reported, and no skin rash or pustules noted at any time, and not on this admission.

The gram stain of the left hip aspirate was reviewed by the ID consultant; it showed no definite organisms but several WBCs were noted with no intracellular organisms seen.

While the ID consultant was still in the microbiology laboratory, the culture plates were reviewed, confirming the following findings: there was no growth on blood agar (BA) and MacConkey (Mac) plates, but growth on the chocolate agar (CBA) plate. The antinuclear antibody (ANA) and rheumatoid factor (RA) tests were negative.

• *How does this new information help narrow the differential diagnoses, even without more data at this time?*

• *What additional laboratory tests would you request or expect the laboratory to do?*

The crystal exam of the left hip aspirate of 9/7/15 had not yet been reported but it turned out later that day to be negative for gout (as reviewed with the pathologist).

Figure 13.4b Two-day-old culture of the left hip aspirate on CBA: growth on 9/9/15. The plate shows typical colony morphology (form) of this organism: small (0.5–1.0 mm in diameter), raised, and glistening.

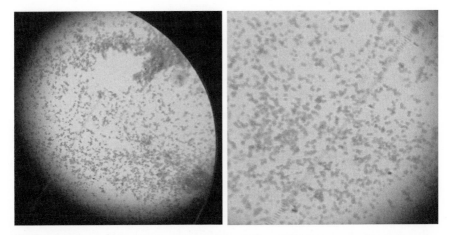

Figure 13.4c Gram stain of 2-day-old growth on CBA on 9/9/15: gram-negative diplococci and coccobacillary forms. (*See insert for color information*)

The 2-day-old culture on the CBA plate reviewed on 9/9/15 is shown in Fig. 13.4b and the gram stain from the CBA plate growth is shown in Fig. 13.4c. The gram stain from the CBA showed gram-negative coccobacillary forms, kidney bean-shaped organisms that were oxidase positive (Fig. 13.4d), and fermented glucose.

Figure 13.4d Oxidase test on filter paper done on the CBA positive culture isolate on 9/9/15: the test was positive (dark purplish color).

Additional biochemical reactions were positive for the following: glucose (GLU), nitrite (NO_2), and peptidase [L-propyl-β-naphthylamide] (PRO), as shown in Fig. 13.4e, confirming the identification of the organism, which was beta-lactamase negative.

Modifications in the antimicrobial therapy were made on 9/9/15 following the visit to the microbiology laboratory.

- *What is your final diagnosis or organism identification?*
- *What antimicrobial agent would you now recommend for this patient?*

The culture was reported positive for *Neisseria gonorrhoeae*.

The working diagnosis was revised to migratory polyarthralgia or polyarthritis; the cause of the arthritis was deemed to be *Neisseria gonorrhoeae*. The gonococcal arthritis of the left hip was thought to have resulted from disseminated gonococcemia.

HIV serology (screen) was negative.

The antimicrobial treatment was thus changed on 9/9/15 to ceftriaxone 1 g daily. She was also given 1 g of oral azithromycin, once only, to cover for any untreated chlamydia, and vancomycin was discontinued.

Additional joint aspirations were recommended.

She was discharged to outpatient follow-up 2 days later on 9/11/15.

Outpatient follow-up, complications, and readmissions

This patient had three admissions to the hospital. The first admission was from 9/6 to 9/11/15 (5 days). She was readmitted twice more, from 9/14 to 9/20/15

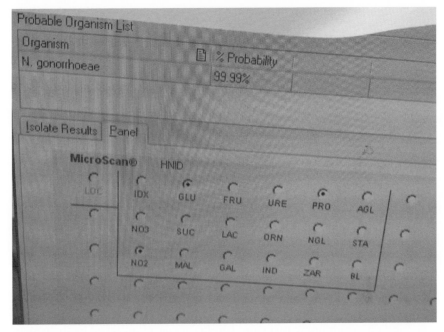

Figure 13.4e Biochemical reactions of the CBA culture of the left hip, as noted on 9/9/15.

and again from 10/5 to 10/8/15. These admissions were primarily for pain control of the left hip.

Overall, she had ultrasound-guided aspiration of the left hip done on three occasions. The initial aspirate on 9/7/15 yielded 18 mL of cloudy greenish-yellow fluid that was obviously purulent. That culture was positive for *Neisseria gonorrhoeae*. The second aspirate was on 9/9/15 and yielded only 2 mL of cloudy pinkish fluid that did not grow in culture. The last aspirate was on 9/14/15 during the second admission: 4 mL of turbid blood-tinged fluid was aspirated that yielded a negative culture.

Markers of inflammation followed over a 5-week period (ESR and CRP) are noted in Tables 13.4b and 13.4c. These markers improved as the patient progressively improved clinically in her overall well-being and ambulation.

Table 13.4b Serial C-reactive protein (CRP) values measured over time.

Test/ date	Reference range	9/6/15	9/8/15	9/16/15	9/19/15	9/30/15	10/5/15	10/5/15	10/12/15
CRP	Latest range: 0.00–0.30 mg/dL	12.90	7.99	7.83	1.58	1.40	1.62	1.95	1.01

Table 13.4c Serial erythrocyte sedimentation rate (ESR) values measured in the patient.

Test/ date	Reference range	9/6/15	9/7/15	9/8/15	9/10/15	9/16/15	9/19/15	9/30/15	10/5/15	10/12/15
ESR	Latest range: 0–20 mm/ hour	117	120	110	85	70	75	42	39	36

The patient made multiple visits to the EDs of two hospitals, nine altogether. Four of these visits were made before she was finally admitted on 9/6/15. Three of the visits led to the three admissions and in two ED visits after the initial admission on 9/6/15, she was managed symptomatically without admission to the hospital. The visit on 9/29/15 was for constipation and urinary retention, thought to be due to narcotic pain medication (oxycodone-acetaminophen) and systemic diphenhydramine (over-the-counter medication) taken by the patient. The other visits were for joint pain complaints and difficulty ambulating.

The patient was seen in the outpatient setting by the primary care obstetrician, as well as beingfollowed by the ID physician. She was seen in my office on three occasions: 10/9/15, 10/1/15, and 9/25/15. She was seen and followed in the hospital during the two admissions in September, 2015.

Ceftriaxone antibiotic treatment was extended for 3 weeks from 9/9/15 until 10/2/15, as we followed the clinical parameters and laboratory markers (ESR and CRP). The initial plan had been for a 14-day treatment, but this was extended because of the slow recovery and prolonged pain.

During the last office visit on 10/9/15, she was 25.5 weeks pregnant; she was ambulating close to normal, and no longer needed a walker. She has continued to do well since.

The recovery was prolonged because the diagnosis was delayed, and consequently also the initiation of specific and appropriate antimicrobial therapy.

During the initial hospital admission (9/6–9/11/15), she was seen by various consultants, including rheumatology, orthopedic, neurology, obstetrics, and infectious diseases. The left hip aspirate procedures were done through ultrasound guidance by the radiologists.

Case discussion
Our patient had disseminated gonococcal infection (DGI), presenting as left hip septic arthritis at the time of admission and confirmed by positive culture of the aspirate.

The diagnosis was not immediately apparent, and was delayed for more than 2 weeks because the symptoms were confusing, initially mimicking gouty arthropathy, especially with involvement of a typical joint for gout (first MTP joint). Because of her pregnancy, there was also reluctance to undertake invasive procedures, or any diagnostic contrast examination in her work-up. She ended up with multiple visits to EDs before she was finally admitted, with enough symptoms to warrant a drainage procedure. Without the left hip aspirate and culture, we might not have been able to make the correct diagnosis.

In retrospect, she had migratory polyarthralgia or polyarthritis, with initial involvement of the left big toe MTP joint, then the left knee, and finally the left hip. In the setting of a young healthy pregnant woman with asymmetric symptoms, DGI should have been on the list of differential diagnoses.

Infection with *Neisseria gonorrhoeae* remains an important world-wide problem. In the USA, it has been shown to be the second most prevalent sexually transmitted infection (STI), as well as the second most commonly reported communicable disease [1,2].

Although infections with *N. gonorrhoeae* remain an important disease, DGI occurs in only 0.5–3% of patients infected with this organism. Most affected persons are younger than 40 years, and more women are affected than men in most studies [3].

There was no recent history of symptomatic genital infection or STI in this patient but this fact is not unusual, but rather typical in patients with DGI [3]. Our patient had no recall or signs of any recent STI, and had not been sexually active since her pregnancy (for at least 4 months).

Patients with DGI typically present with one of two syndromes [3]:
• tenosynovitis, dermatitis, and polyarthralgias without purulent arthritis, or
• purulent arthritis without associated skin lesions.

Our patient had purulent left hip arthritis with no skin lesions (also no rash or pustules), and no fever, chills, or other systemic symptoms other than severe left hip pain. This lack of systemic symptoms contributed to the initial confusion and subsequent delay in diagnosis.

The gram stain of the hip aspirate did not show any organisms, but the culture was positive, confirming the diagnosis. If she had received a *N. gonorrhoeae*-active antimicrobial agent, the culture could have been negative. The diagnosis may have been missed, unless we used non-cultural methods like nucleic acid amplification testing (NAAT) [3,4]. Luckily, she received clindamycin, which has no activity against *Neisseria* organisms.

The laboratory culture and stain results were typical for *Neisseria* organisms in the setting of DGI. The culture was positive on CBA, but negative on BA and Mac [4]. The cultural characteristics were also typical, with small, raised, glistening colonies (see Fig. 13.4b) and no intracellular organisms seen on the gram stain, as would be the case in urethral smears [4].

The isolated *N. gonorrhoeae* was penicillin sensitive (beta-lactamase negative), as is typical for agents causing DGI, agents that are typically classified as "serum resistant" [2,4]. The patient was therefore appropriately treated with ceftriaxone.

She responded more slowly than would be expected because of the delay in making a diagnosis and starting specific treatment.

Lessons learned from this case

- The diagnosis of DGI can be very difficult in a patient without systemic symptoms, as was the case in this patient.
- A high index of suspicion is needed in the sexually active young person (especially a pregnant female) presenting with acute polyarthritis, polyarthralgias, or oligoarthritis.
- Lack of recent sexual activity or STI should not dissuade one from thinking of DGI in the appropriate setting.
- Lack of systemic symptoms should not dissuade one from thinking of DGI. This is quite typical in the "purulent arthritis" variety of DGI.
- Many consultants were involved initially because the diagnosis was not clear.
- The multiple visits to the ED were a reflection of the difficulty in reaching a diagnosis.
- As is good medical practice, appropriate cultures should be obtained or attempted before beginning antimicrobial therapy in non-life-threatening conditions.
- Good laboratory support is needed for the timely diagnosis of *Neisseria gonorrhoeae* in the patient with DGI.

References

1 CDC. Gonococcal infections in adolescents and adults. Available at: www.cdc.gov/std/tg2015/gonorrhea.htm, accessed March 4, 2016.

2 Price GA, Bash MC. Epidemiology and pathogenesis of Neisseria gonorrhoeae. Available at: www.uptodate.com/contents/epidemiology-and-pathogenesis-of-neisseria-gonorrhoeae-infection?topicKey=ID%2F7587&elapsedTimeMs=7&source=search_result&searchTerm=epidemiology+and+pathogenesis+of+Neisseria&selectedTitle=1%7E150&view=print&displayedView=full, accessed March 4, 2016.

3 Goldenberg DL, Sexton DJ. Disseminated gonococcal infection. Available at: www.uptodate.com/contents/disseminated-gonococcal-infection?topicKey=ID%2F7603&elapsedTimeMs=10&source=search_result&searchTerm=disseminated+gonococcal+infection&selectedTitle=1%7E33&view=print&displayedView=full, accessed March 4, 2016.

4 Janda WM, Gaydos CA. Neisseria. In: Murray PR, Baron EJ, Jorgensen JH, Landry ML, Pfaller MA (eds) *Manual of Clinical Microbiology*, 9th edn. Washington, DC: ASM Press, 2007, pp. 601–20.

Index

Cases in Clinical Infectious Disease Practice: Obtaining a Good History from the Patient Remains the Cornerstone of an Accurate Clinical Diagnosis: Lessons Learned in Many Years of Clinical Practice, First Edition. Okechukwu Ekenna.
© 2016 John Wiley & Sons, Inc. Published 2016 by John Wiley & Sons, Inc.